MODERN WORLD NATIONS

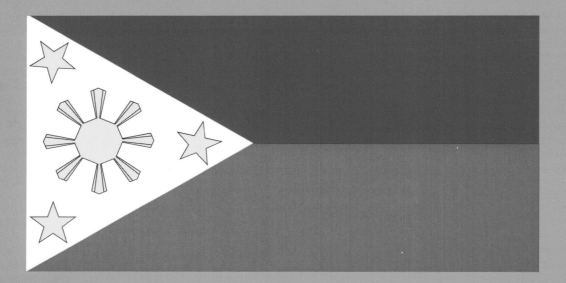

The Philippines

Tammy Mildenstein

and

Samuel Cord Stier

Series Consulting Editor
Charles F. Gritzner
South Dakota State University

A Haights Cross Communications Company

Philadelphia

Frontispiece: Flag of The Philippines

Cover: Harbor and village in the Philippines.

CHELSEA HOUSE PUBLISHERS

VP, NEW PRODUCT DEVELOPMENT Sally Cheney
DIRECTOR OF PRODUCTION Kim Shinners
CREATIVE MANAGER Takeshi Takahashi
MANUFACTURING MANAGER Diann Grasse

Staff for THE PHILIPPINES

EXECUTIVE EDITOR Lee Marcott
PRODUCTION EDITOR Noelle Nardone
SERIES DESIGNER Takeshi Takahashi
COVER DESIGNER Keith Trego
PHOTO RESEARCH 21st Century Publishing and Communications, Inc.
LAYOUT 21st Century Publishing and Communications, Inc.

A Haights Cross Communications ◀ Company

http://www.chelseahouse.com

First Printing

1 3 5 7 9 8 6 4 2

Library of Congress Cataloging-in-Publication Data

Mildenstein, Tammy, 1968-
  Philippines / Tammy Mildenstein and Samuel Cord Stier.
     p. cm. -- (Modern world nations)
  Includes bibliographical references and index.
  ISBN 0-7910-8024-2
  1. Philippines--Juvenile literature. I. Stier, Samuel Cord. II. Title. III. Series.
  DS655.M55 2004
  959.9--dc22

                                                          2004013693

Table of Contents

MODERN WORLD NATIONS

The Philippines

1

The Philippines: Country of Many Islands

C an you imagine what it would be like to live on an island, surrounded by the lapping blue waves of the Pacific? What if you lived in a country made up of *thousands* of islands, spreading out to the horizon? All your friends and family would live on islands, some on your island and others on neighboring islands or islands located several days away by boat.

A country of islands brings up many puzzling questions. What would make these islands into a country at all? Wouldn't each island feel more like its own community than part of a bigger country? How would you travel? Would you have to own your own boat? How would you talk with other people on other islands? Do families live on the same island? If not, how do they keep in touch with family members on different islands? How would the government have any control over so many islands?

The Philippines is a great example of the country you are imagining. It is a country spread across an estimated 7,107 different islands. The way the Filipinos handle all of these questions of day-to-day life on islands is one thing that makes the Philippines unique. Throughout this book you will see the special role that islands have played in the Philippines.

LOCATION IN THE WORLD

If you lived in the Philippines, you would be on the opposite side of the world from the United States, in a region called Southeast Asia. Your closest neighboring countries would be China, Vietnam, Indonesia, Thailand, and Malaysia. The islands that make up the country of the Philippines are scattered over 780,000 square miles (2 million square kilometers) of the South China Sea and the Pacific Ocean. This total area is about the size of Arizona, New Mexico, Utah, Nevada, Oregon, and California combined. If you pushed all of the Philippine islands together like puzzle pieces, though, the total land area would be the size of only one state, say Arizona or New Mexico. In this way, the Philippines is both large and small at the same time.

Also, being a country of islands means that the Philippines is both near the Asian mainland and separate from it at the same time. The Philippine borders do not touch any other countries, and so the country is more affected by people who visit the islands than by its mainland Asian neighbors. In this way, the people of the Philippines are a cultural mix of both Asian and more-distant cultures.

The Philippine islands historically have offered an ideal resting place for immigrants looking for new lands to settle and for sea-faring explorers seeking new lands for their countries. Over the years, the Philippines has become host to many different cultural immigrants. Today, the country truly is a melting pot of cultures. If you lived there, you would notice that most of the Filipinos are from Asian cultures, including Malay, Chinese, Papuan, Indian, and Japanese. It might confuse you that there also was a lot about the

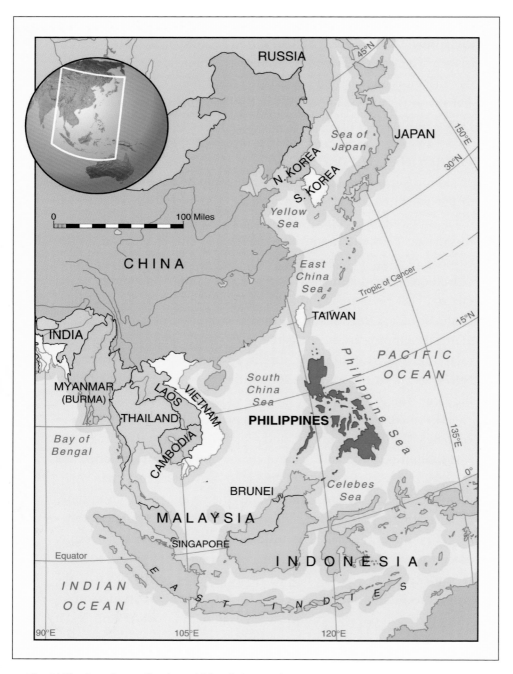

The Philippines is a collection of islands in Southeast Asia, spanning 780,000 square miles (2 million square kilometers) of the Pacific Ocean and the South China Sea.

Philippines that did not seem Asian at all. In every neighborhood or village, you are almost certain to see people playing basketball! Also, in addition to speaking local languages, people know English. There are also Filipino words like *kumusta, peso,* and *Lunes,* which sound a lot like Spanish. England, the United States, and Spain are not even close to the Philippines, so how did think English and Spanish words became so popular in the Filipino culture?

SPANISH COLONIZATION

Because of the Philippines' isolated, but central, location in Southeast Asia, it was a strategic choice for imperial countries (those that sought to govern other countries or territories in order to use their natural resources) from the West that sought a base for trade with Asia. Explorers from Spain were the first Western imperialists to land in the Philippines and to claim these islands as a colony of their own. Much of the Spanish culture remains in the Philippines today. When comparing the Philippines with other Southeast Asian countries, it is hard not to notice the distinctive Latin feeling that comes from the country having been a Spanish colony. This "Latin" (or Spanish) quality is evident especially in the language, which has a lot of Spanish mixed into the Filipino dialects. There are Spanish words for numbers, the time of day, and virtually everything in the kitchen.

Spanish influences are present in other aspects of Filipino culture as well. Most Filipinos practice Catholicism, which was introduced by the Spanish. Also, if you lived there, some of your favorite things to do would come from the Spanish. You might enjoy playing guitar, going to neighborhood parties called *fiestas,* or watching cock fights (a fight between roosters).

AMERICAN INFLUENCE

After the Spanish, a second Western country recognized the strategic location of the Philippines in Asia and decided to

Many imperialist countries targeted the Philippines because of its strategic location in Southeast Asia. Spain was the first country to colonize the Philippines, and its influence is still very evident today. For examples, most Filipinos practice Catholicism, introduced by the Spanish, and attend mass in churches like the one seen here.

move in. In 1898, the United States fought Spain and won, thereby asserting its right also to colonize the Philippines. With this new colonial power came a new set of Western influences. The Americans brought ideas of democracy, schools, roads, and electricity to the Philippines. English came with the Americans as well. This gave the Filipinos a common language that enabled them to communicate with one another for the first time. English was also a powerful tool for understanding the Americans, as well as for interacting internationally. Today, one of the most striking things about the Philippines is the fact that almost everybody speaks at least some English.

Although the Philippines has been an independent nation for more than 50 years, it maintains very close ties with the United States. This relationship has at its foundation the years of partnership between the two countries. Filipinos fought on the American side in World War II and the Vietnam War. In exchange, Filipinos received green cards for immigration to the United States. Many families benefit greatly from having family members in the United States, and many households rely heavily on the regular income they receive from overseas relatives.

Americans, on the other hand, have looked to the Philippines as a strategic location for military outposts. Up until 1991, the Philippines harbored the largest American overseas naval and air force bases, housing more than 40,000 permanent military employees. Although these facilities were closed in the early 1990s, there are still U.S. military operations in the Philippines. They are temporary arrangements, however, and on a much smaller scale. A second major relationship between the United States and the Philippines is a more commercial arrangement. Many American businesses have factories in the Philippines and rely on Filipino factory workers to make their products at a lower cost.

CULTURAL DIVERSITY

You probably sense that the Philippines is a country of great diversity with surprising combinations of different cultures. Although it operates as a single country, there are many different cultural communities on the islands. These peoples are as diverse as the island areas where they live. There are islands where people grow their own food, but a few miles away other people buy all their food in grocery stores. On some islands, businesspeople take taxis, wear suits, and work in tall office buildings; just outside the city there are people wearing hand-woven cloth and growing rice in watery, terraced fields that reach to the sky.

It is not surprising that every island has fishing villages near its beaches. The men in these villages fish for a living, and almost every family has its own fishing boat. When you visit a fishing village in the Philippines, the first thing you notice is that the beaches look like parking lots for hundreds of brightly colored boats! The families in the fishing villages often live in houses made of bamboo and woven leaves. Some fishing villages are built entirely above the water, with every house on wooden stilts.

You would think that in a country of islands, everybody would fish and live by the sea. On some of the largest islands, however, there are high mountains. The mountain tribes that live there may never have seen the ocean before, even though it is only a few dozen miles from their homes. The people of the mountains do not fish; rather, they grow rice and vegetables, terracing the hillsides into giant green staircases of farm fields.

In the far north, some islands lie closer to Taiwan than to the rest of the Philippines. On these islands the typhoon (hurricane-like storm) winds can be so strong that they are measured by the size of the animal that they would blow over. If the weather forecast is a "chicken typhoon," it is a good idea to bring the laundry (and the chickens!) inside. If a "water buffalo" typhoon is on the way, families bring in all their belongings and pray that their houses will survive the storm. In fact, unlike the bamboo houses of Filipino fishing villages, the houses on these northern islands have walls of stone almost three feet (one meter) thick—built to withstand the howling winds of these seasonal storms!

The southern islands of the Philippines are where many people practice Islam rather than Catholicism. In those areas, there are many mosques instead of churches. The cultural communities in this region are perhaps more similar to the neighboring Southeast Asian Muslim countries, Indonesia and Malaysia.

ECOLOGICAL DIVERSITY

In addition to being culturally diverse, the Philippines is known for its great wealth of life forms (animals and plants), sometimes called biological diversity or biodiversity. In the Philippines you can see the world's largest bats, as big as eagles with six-foot (two-meter) wingspans! They also have the world's smallest buffalo, which is about the size of a big dog. There are wonderfully colorful birds of all shapes and sizes. Hornbills are birds with big horny beaks that are used to eat forest fruits and that make hornbills' calls sound as if they have stuffed noses. Some of the Philippine forest frogs are so well camouflaged that you can be staring right at one and think it's just a leaf.

The Philippines does not have big deadly animals like lions or grizzly bears. Instead, Filipinos walking in the forest tend to watch out for deadly snakes like the aggressive king cobra, which grows up to 30 feet (9 meters) long and stands up to 6 feet (2 meters) tall. On the other end of the deadly snake spectrum is the bamboo viper. This snake is only a foot (30 centimeters) long, but if it bites you, you may only have seconds to live before the venom kills you. For this reason, Filipinos call it the "two-second" snake.

During the period of time when the plants and animals on the islands were developing, most of the islands were separate from one another. This meant that unique species of plants and animals, like the ones listed above, were forming on islands separated by only a few miles. After years of separation and the development of many different life forms, the Philippines is one of the smallest areas in the world where you can see many different kinds of animals and plants.

We will explore these and many other aspects of the Philippines in the rest of the book, so take a moment, close your eyes, travel to the other side of the world in your mind, and *Mabuhay sa Pilipinas*! (Welcome to the Philippines!)

Natural
Environment

If you want to understand a country's people, it is important to know something about natural environments in which they live, work, and play. How could anyone understand Eskimos (Inuit), cowboys, farmers, and surfer dudes living in the United States unless he or she knew how big and diverse the country's environment was? In the same way, if we want to understand the Philippines, it is important to know about the natural environment there.

It was mentioned already that Filipinos live on thousands and thousands of islands. In this chapter, you will explore where these islands are, where they came from, and what the land is like on them. You will look at how Filipinos have adapted to live on these islands, how they use the environment in day-to-day life, and, in the process, how their activities have changed their natural world.

LOCATION

Where the Philippines is located affects everything, including the economy, culture, origins of its people and, of course, its environmental conditions. If you find California on a globe and trace your finger west across the Pacific Ocean, the first land you hit is Asia. If you look in the water south of China and Japan, east of Vietnam, and north of Indonesia, you will find a small cluster of islands that makes up the Philippines. Some Filipinos say that this cluster of islands looks like a man squatting in a field of water to plant his rice.

The Philippine archipelago (island chain) is located in the area called Southeast Asia. Other countries in this region are Vietnam, Thailand, Laos, Myanmar, and Malaysia on the Asian mainland, and the archipelago of Indonesia. In this area, where the Pacific Ocean meets the South China Sea, lie the islands of the Philippines. Because of its unprotected insular (island) nature, the Philippines is subjected to more severe weather patterns than are its mainland neighbors. Also, being located in a large water body between two big mainlands (Asia and Australia) makes the Philippines an important stopover for migrating birds and marine mammals.

LANDFORMS

On a map, the Philippine islands all look pretty similar to each other. They have different shapes and sizes, of course, but you would guess that they would all look alike if you visited them. Actually, there is an amazing amount of diversity among these islands. Some islands are surrounded by sandy beaches, and others have rocky limestone edges. The islands range from fairly flat and covered with palm trees to having very high mountain peaks covered with pine trees. The differences in the islands' interiors can be attributed to how the islands were formed and the movement of the islands as parts of great landmasses.

Island Origins

There has been a lot of scientific investigation into the formation of the Philippine islands. Scientists have used rocks and soils

Though they span a large area of the Pacific Ocean, the Philippine islands, located east of Vietnam and north of Indonesia, are actually just over 115,000 square miles (298,000 square kilometers) in size. Because of their location, the islands feature a lot of diversity, from the hundreds of animal species to the landscapes that include both flat beaches and high mountain peaks.

on the islands and the ocean floor around them to identify their origins. Some of the islands broke off of mainland Asia long ago and then moved into their present location. Other islands originated from hot volcanic magma (molten rock) built up from the ocean floor. The parts of these islands above water are just the tops of volcanoes. (The Hawaiian islands were also formed in this way). Finally, other islands are made from coral atolls (land made of coral reefs that once surrounded a volcano). The soil on these islands is made from rocky limestone that used to be part of coral reefs in the ocean. These underwater reefs were pushed up out of the ocean and began to break down into flat islands that became homes for plants, animals, and people.

Plate Tectonics

The old pieces of mainland Asia, as well as the volcanic and limestone islands, are all jumbled together thanks to a process called plate tectonics. Beneath the continents and oceans that we see on the surface, Earth's crust is about 44 miles (70 kilometers) thick and made up of many separate plates (flat pieces), and just like students, Earth is restless! These plates move around slowly because of thermal changes deep within Earth's core, and, as they move, they rub against each other. The friction between plates is responsible for earthquakes, volcanoes, and the creation of mountains. In fact, most of the world's mountain ranges are the result of plates crashing together and forcing land upward. In the United States, the Rocky Mountains are a good example of this. These mountains were formed by the Pacific plate and the North American plate colliding and over time forcing large landmasses into the air. The earthquakes in California are the results of plate tectonics, too. This is where the Pacific plate rubs the North American plate. You can imagine these huge plates constantly forcing themselves into each other under the ocean. When the pressure finally gets too strong, something has to give, and . . . earthquake!

In the Philippines, there are three major plates at work. The Eurasian plate, which is under mainland Asia, and the Pacific plate, under the Pacific Ocean, are the two largest plates. Between these two huge moving masses of earth is the smaller Philippine plate. The plates' movements into each other are the forces that created the volcanoes, pushed up coral reefs, and broke off pieces of mainland Asia that are now the Philippine islands. There are islands on each of these plates, which means that, as the plates underneath move, the islands move around, too. Because this occurs very slowly, we do not notice the movement when looking at them. Scientists know, however, that the islands were arranged differently millions of years ago.

Landforms on the Islands

As was explained earlier, the islands of the Philippines have three distinct origins: separated pieces of mainland Asia, volcanoes, and ancient coral reefs forced up into limestone atolls. Some islands were formed by several smaller islands of different origins being pushed together over time. If you visit any of the Philippine islands, there are clues you can look for to guess how the island was formed.

Volcanic Features

Volcanic islands may have steep, pointy volcanic peaks, or mountains, that were once volcanic but have broken down over time. Many of these volcanoes are dormant, or inactive, but some, like Mount Mayon in the Bicol region, in southeastern Luzon Island, are still active. At night, in the towns at the foot of Mount Mayon, you can see red-hot lava oozing down the sides of this amazing volcano.

There are 37 volcanoes in the Philippines, 18 of which are considered active. Sometimes, a dormant volcano takes all of its neighbors by surprise and, after hundreds of years of inactivity, blows its top off without warning. A famous example of this was the violent explosion of Mount Pinatubo in 1991. This

Portions of the Philippine islands are made of the tops of volcanoes, many of which are still active. Mount Mayon, seen here, in Legaspi, is the most volatile volcano in the archipelago and has erupted as recently as 1993 and 2000.

volcano was considered virtually extinct (dead) after 600 years of complete dormancy when it erupted, shooting ash, mud, and rocks up to 25 miles (40 kilometers) in the air. The heavy *lahar* (ash, mud, and rock mixture) covered towns miles away, burying many buildings with as much as 10 feet (3 meters) of ash. People had to crawl out on their roofs to escape their buried homes, and, because all the cars and roads were buried, there was no way for people to get away from the disaster. More than 900 people were killed, and thousands of people lost everything they owned.

Gray ash from the explosion fell on Philippine islands far from the blast. Filipinos in these areas described waking up to the dusty whitish blanket and thinking it was Christmas, because it looked like the snowy Christmas scenes in the movies.

Pinatubo's surprise is still a reality in the Philippines. Filipinos have rebuilt their houses six feet (two meters) above where the old houses were. Some of the old (and taller) buildings are still being used, but people have to enter through what used to be second story windows. Filipinos have even found a way to use all the excess ash: They make things like cement, as well as ceramics.

Coral Islands

Islands that were once coral reefs have limestone features that are very distinctive. Some have steep limestone cliffs that plunge into the ocean, like on Coron and Mindoro islands. Others islands have odd hills made of limestone that have since been worn down by weather. Bohol Island is famous for its Chocolate Hills, which make a bizarre landscape of round limestone hills that are brown from the burning of dried vegetation to improve livestock grazing. Most coral islands, however, are very flat and only a few feet above sea level. Sometimes, if you look closely at the exposed limestone areas on any of the Philippine islands, you can find fossils of the marine animals that lived in and built the coral reefs.

Separated Pieces of Mainlands

On the islands that broke off from mainland Asia, there usually are leftover signs of the Asian continent. These islands or island areas have very different soils and plant communities. They are much more similar to mainland Asia's soils and plants than to those of their island neighbors. A good example of this is Luzon, where the northern part of the island is a piece of Asia that fused with the rest of the island millions of years ago. This part of Luzon has very different soil and plants from the rest of the island. Palawan Island is another example: It once was attached to Borneo and a number of other islands in the south. It has plants and animals that are unique to the Philippines but common to Borneo. They are so unusual that they are hard to name: bearcats (not really a bear or a cat), peacock pheasants (not really a peacock or a pheasant), and mouse deer (not really a mouse or a deer).

Interior Mountains

There are some similarities in landform among all the Philippine islands, no matter what their origin. All of the islands are surrounded by ocean, thereby creating beach areas on the perimeter where people can access the water with boats. Many of the Philippine islands have higher mountain ranges in the centers. Generally, the larger the island, the more numerous and higher the mountains. In fact, the two highest points in the country are found in mountains on the two largest islands. Mount Pulog on Luzon Island rises to 9,613 feet (2,930 meters) from within Central Cordillera mountain range. Mount Apo on Mindanao Island reaches a spectacular height of 9,692 feet (2,954 meters), rising straight out of the sea.

WATER FEATURES

Water plays a big role in the lives of most Filipinos. This is because, as a country of islands, they are surrounded by it and also because Filipinos need freshwater sources for their day-to-day lives. Most towns and cities in the Philippines are located near water, and there are a number of important water features in the country worth mentioning.

Saltwater Features

Natural saltwater areas, like the ocean and its bays, offer access to seafood. Filipinos gather fish, mollusks, shrimp, and other seafood. They have learned to "farm" fish using enclosures in saltwater areas as pens. Many Filipino communities are located on the coast, bays, and inlets for easy access to these stores of food. Laguna de Bay, which is a very big inlet on Luzon Island, is home to millions of people and is the source of much of the fish that is eaten in this region.

Sulu Sea

Sulu Sea is the wide, open sea between Palawan Island, Borneo, and the Philippines. This area is one of the most dangerous travel routes for ships because of the pirates that stalk this sea looking for ships to loot. Despite this danger, Sulu Sea is used as a fishing area for Filipinos. This area is probably best known for some of its natural qualities. Sulu Sea is so vast that it holds a lot of natural marine life, including a great variety of coral and many endangered species of marine animals.

Bais Strait

Bais Strait is another important water feature for wildlife. Between Negros and Cebu islands, Bais Strait is a deep channel with swift currents. It is an important migratory route for marine mammals. If you ride a pump boat (a wooden boat with bamboo stabilizers) through this area, it usually will stir up dolphins, porpoises, and whales. The dolphins and spinner whales get so excited that they play games and do crazy tricks while swimming and chasing each other around in front of your boat!

Freshwater Features

Freshwater sources are also important to Filipinos, because finding clean drinking water can often be a struggle, especially during the dry season. Freshwater also is important for agricultural irrigation. Broad, fertile river valleys can be found between the mountains in the interior of many islands. These river valleys usually are lined with villages and bright green rice fields. They are the source of most of the rice, fruits, and other foods eaten on that and often on neighboring islands. Rivers are also important to people in other ways. Women can be seen washing their family's clothing at the river while children use the river to play in and cool off from the summer heat. Once again, the longest rivers in the country are found on the largest islands. The Cagayan, Pampanga, and Agno rivers on are Luzon Island, and the Rio Grande de Mindanao and the Agusan River are on Mindanao Island.

Agusan Marsh

Because it is one of the largest freshwater marshes in Southeast Asia, Agusan Marsh is a well-known freshwater feature. It is the most important stopover point for migratory birds in all of Asia. More than 102 species of migratory birds come here from as far away as China, Russia, and Japan to escape the harsh winters. The Manobo people have permanent floating houses built on the marsh, and the marsh provides virtually everything the Manobo need to live.

ECOSYSTEMS AND PLANT AND ANIMAL LIFE

Biodiversity is a term that comes from two words—"biological" and "diversity"—and refers to the number of different plant and animal types in a place. The Philippines has incredibly high biodiversity for three reasons: First, it is in the tropics, where the numbers of plants and animals tends to be very high; second, it is an archipelago, which means that there are many different islands with many different ecosystems; finally, it has been isolated from the mainland areas for a long time. As an archipelago, the Philippines has no direct contact with other countries. Because the islands have been isolated for a long time, they have developed many species of plants and animals not found elsewhere.

Animal and plant species that have developed in an area and are unique to that area are said to be endemic to that area. The Philippines has one of the highest numbers of endemic species in the world, mostly because of its separation from neighboring countries. You may think that, because the Philippines are thousands of separated islands, there must be endemic plant and animal life on individual islands. You would be right about islands contributing to the large number of endemic species in the Philippines, because there are island endemics, but the islands are not as disconnected as you might think, because of the changes in sea level not too long ago. During the last ice age, the oceans were several hundred feet lower and many islands were connected. Because of this, these islands share similar plants and animals. (For additional information on Philippine wildlife, enter "Philippines animal species" into any search engine and browse the sites listed.)

Marine Environments
Coral Reefs

Coral is a living, growing organism that forms structures on shallow ocean floors called reefs. Coral reefs are famous around the world for having beautifully colored fish and corals. People like to dive and swim next to coral reefs just to look at them. In the United States, the only places where coral reefs occur are the Florida Keys and Hawaii.

Most of the Philippine islands are surrounded by coral reefs. These structures are such an important habitat for so many fish and other marine life that they are described as forming their own unique type of ecosystem. Scleractinian corals, or reef-forming corals, first appeared on earth more than 200 million years ago and are now the world's primary reef builders. Tubbataha Reef, in the Philippines' Sulu Sea, is one of the biggest reefs in the world, covering an area of 82,037 acres (33,200 hectares). This one reef has more than 300 coral species and 379 species of fish living in it! It is home to a diversity of unique fish with names like the giant bumphead parrotfish, the Napoleon wrasse, the needlefish, and the pink flasher wrasse. Six of the world's eight species of sea turtles can be found in

Coral reefs, famous for their unique and attractive marine life, surround most of the Philippine islands. This Gorgonian sea fan coral is one of many species that make up the reefs around the Philippines.

Sulu. There are also eight species of whales and dolphins in these waters. Huge sea cows called dugongs also live there, as do ferocious and often deadly saltwater crocodiles.

Filipinos have developed many uses for their reefs. The reefs provide a useful source of fish for eating, and many fisher-men take their boats to the nearest reef to catch fish for their families and to sell in the market. Reefs also have been used as gravel or in making cement by breaking up the living coral and bringing it to shore. Both of these uses have led to a sharp decrease in the amount of living coral reefs in the Philippines. Fishing used to be practiced with small nets or hooks and lines, which did not have a negative impact on the coral reefs.

Recently, however, fishermen interested in getting more fish in easier ways have developed more efficient fishing techniques. Now some fishermen use two methods, dynamite and cyanide poison. Both kill the coral and many other fish in the process, and many fishermen complain of how few reef fish are left for fishing. Because of this destruction, there is a lot of pressure for the Philippines to protect the coral reefs.

Sea-grass Beds

Another type of marine ecosystem in Philippine waters is sea-grass beds. These are found in shallow waters near the island shores, and, as the name suggests, they look like big grassy fields underwater. These areas hold far fewer plant and animal species than do the bustling reefs, but what lives there usually lives only in that place, so it is a special place. The types of animals you can find in sea-grass beds include sea cows (dugongs), sea horses, pipefish, and knifefish.

Filipinos use sea-grass beds to make a very interesting ingredient called carrageenan. Carrageenan is a thickening substance that is often used in place of gelatin and is becoming more popular in the American food industry. Next time you eat a candy bar, jelly beans, gummy bears, or any product that needs a gelling agent, look in the ingredients for carrageenan. If it is there, you just might be eating sea grass from the Philippines.

Deep-sea Areas

Farther out from the islands, away from the sea-grass beds and the coral reefs, the sea gets much, much deeper. In fact, the third-deepest place in the global sea is located right in the Philippines. The Philippine Trench is located just east of Mindanao Island in the Pacific Ocean. The trench descends to a depth of 34,580 feet (10,540 meters), which is deeper than the Grand Canyon. If you dropped a two-pound (one-kilogram) metal ball into the water above the Philippine Trench, it would take more than an hour for it to reach the bottom!

The deep seas are where Filipino fishermen find big fish like sharks, tuna, dorado, and barracuda. They tend to fish for these using big nets that they throw out from their boat in the evening and then pull in during the early morning hours. In recent years, deep sea fishing success has taken the same turn as coral reef fishing. As the number of people has increased, fishing needs have as well. As a result, there has been a notice-able reduction in the numbers and sizes of fish caught in the deep sea areas. Fishermen say that they have to work much harder to catch the amount of fish that they caught even 10 years ago. They have to travel much farther out to sea and fish much longer just to bring in enough fish to make a living. Overfishing leading to reduced fish populations is a common problem throughout the world, forcing many countries, includ-ing the Philippines, to consider regulation of fishing in order to conserve this valuable resource.

Mangroves

There is a unique type of ecosystem between the land and the sea that belongs to both marine and terrestrial (land) environments. Mangroves are trees that grow in the muddy, salty beach areas where no other trees can grow. They have adaptations that are roots, called pneumatophores, which emerge just above the level of high tide so the tree can breathe. Another adaptation is that their leaves can excrete salt. They are able to take in seawater without having to swallow ocean salt. Some animals, like fruit bats, visit mangroves just to lick the salt off their leaves. Mangroves also create an important estuary (where the tide meets a river current) habitat for sea creatures like shrimp, crabs, and other crustaceans. Estuaries provide a marine environment in which the young shellfish are protected from harsh waves and large predatory fish.

For several reasons, mangroves are one of the most endangered ecosystems in the Philippines and throughout the world's tropics. They are located right on the beach, where

One unique aspect of Filipino wildlife is mangrove forests, which are neither terrestrial nor marine, but found between the beach and the ocean. The trees survive in salty, wet soil by sending up roots that act as breathing tubes above the highest point of the tide.

people like to build houses and fishponds. Also, mangrove trees have very dense wood, which makes great charcoal for cooking. In some mangrove areas, you can see lots of stumps, because people have harvested trees to cook food. Unfortunately, mangroves are also fairly slow-growing, so it takes a while for these trees to grow back.

Terrestrial Environments

One thing that all the islands of the Philippines have in common is that they once were entirely covered by forests. From the ocean edges to the tops of the highest peaks, there were trees, and these forests differed depending on where they grew. Each of the different forests has its own community of plants and animals living in it and its own uses by Filipinos and has faced its own challenges through time.

Lowland Rain Forest

In the lowlands, from the edge of the mangrove and beach areas up to around 3,280 feet (1,000 meters) are the lowland rain forests. These forests are one of the great marvels of the Philippines. The canopy is made from towering trees that reach well over 120 feet (37 meters) tall, with strongly clinging vines linking them together. Other plants grow in every nook and cranny available in the trees above. Meanwhile, the forest floor is alive with ants, termites, beetles, millipedes, monitor lizards, and 20-foot (6-meter) pythons. In the trees, you can see monkeys, parrots, hornbills, and bats with six-foot wingspans. Other than an occasional pig wallow (mud hole), you cannot find a spot in these forests where at least five things aren't living right on top of each other.

Because of the diversity of living things in rain forests, there are a number of ways people have learned to benefit from them. Filipinos use the forests to hunt wild pigs, flying foxes (fruit bats), and monitor lizards. They gather forest products like wood, rattan, and bamboo, which they use to make many things like baskets, mats, and even the walls of their houses. The wood from Philippine rainforest trees is very strong and has been popular in the international markets for fine furniture.

In the last 100 years, wealthy countries like Japan and the United States have imported enormous amounts of wood from lowland Philippine rain forests. Now the Philippines is missing more than 90 percent of its original forest cover, which is a problem for many reasons. Many animals are now left with little or degraded habitat, which is why so many of the Philippine birds and mammals are endangered species. Second, without the forest, many people no longer have access to the forest products that they have used for many years. Deforestation also results in unstable soils on the mountains that are now missing trees. When it rains, these loose soils turn into heavy runoff that muddies Philippine rivers and mud slides that can bury mountain roads. Also, because there are no longer trees to absorb and contain high levels of water, there now are frequent flash floods. These torrents can wipe out whole towns at the base of a mountain.

Upper Montane Forest

Above 1,000 feet (300 meters) on mountain slopes are upper montane forests. These are forests that can handle the climate and soils at higher elevations and have fewer species than the lowland rain forests. They include pine trees as well as heartier broadleaf trees and are home to many unique birds, rodents, and bats. People living in these areas also use the forests for the products they provide, such as wood, rattan, and animals. Some people clear forest areas for agricultural uses by slashing and burning the vegetation and then planting vegetables and rice in the new open area. Despite the long history of these uses, the deforestation levels are not as high in the Philippine upper montane forests as in the lowlands.

Cloud Forest

On the highest part of the mountains are the cloud forests. These are areas where the trees are dwarfed by the thinner air, cooler temperatures, and less fertile soil. A 20-year-old tree that is 3 feet (1 meter) tall here may be from a species that grows taller than 60 feet (20 meters) down the mountain. The Japanese call these dwarfed trees *bonsai*, and Japanese visitors like to collect these stunted trees to bring home with them. Because of the scarcity of resources available in cloud forests, relatively few species of wildlife live there. Of the bird species and rodents that are able to live in these areas, a good example is the huge cloud rat, which grows to the size of a basketball. Native tribes often hunt cloud rats for meat and use their skins as caps or small backpacks.

WEATHER AND CLIMATE

All of the Philippines lies within the tropics. Weather generally is hot and humid all year round. Compared with places in the United States, it would be most like the middle of summer in the Southeast or like much of the year in southern Florida. The range of annual climate variation is minimal, but most of the Philippines has two fairly distinct seasons.

The "dry season" is typically from about January to May, and the "wet season" is June to December. Dry season temperatures reach well into the 90°F (30°C) range. During this season, Filipinos do their best to avoid the intense heat of the sun. They wear hats or carry parasols (umbrellas used to shade the sun), stay inside during the middle of the day, and do outdoor work in the early morning when it is cooler.

Rainy season temperatures are generally cooler, in the 70 to 80°F (21 to 27°C) range, so the heat is not much of a problem. The rain is a different story. In some areas of the Philippines, it can rain more than 3.5 feet (1.1 meters) in a single month! Maine gets this much precipitation in an entire year, and Maine is one of the wetter parts of the United States.

You can imagine all of the problems Filipinos face with this nonstop rain. None of the laundry ever dries, dirt roads turn into impassable muddy trails, and day-to-day life becomes much more difficult. Because of the difficulty of traveling and working during this season, there are lots of delays in projects. Farmers have to schedule their work carefully, because they depend on the rains to flood their fields and grow the rice.

Because rain plays such an important role in the Philippines, we should distinguish between the special rainy seasons and their importance. In the Philippines, there are seasonal typhoons and two seasons of monsoons that sweep through the country every year.

Typhoon Belt and Seasons

You probably have heard of hurricanes before, especially if you live in the coastal southeastern United States or Hawaii. Every year, these powerful ocean storms hit the east coast of the United States and cause all sorts of damage with high winds and drenching rain. Both hurricanes and typhoons are also called cyclones, because they originate in warm waters that fuel the heat engine of the storms. We call the cyclones that hit North and South America "hurricanes." The name for a cyclone that hits Asia is "typhoon," from the Chinese words *tai* and *feng* meaning "great" and "wind." Generally speaking, coastal areas are most affected, because once the storms hit land, they usually dry out and slow down.

In the Philippines, typhoons occur from May to November, during the wet season, and are destructive only in the northern islands. Being in a typhoon is like being in a hurricane, with terrifyingly powerful winds blowing things over. People tend to take cover in the strongest building around, like a school or a church, and pray that their house will not blow away. In many areas, the great destructive typhoons are fairly rare. As with tornados in the United States, people do not do anything to prevent the destruction of these storms, they just hope they do not come. There are also areas in the Philippines, particularly in the islands located farthest to the north, where strong typhoons hit every year. In these areas, Filipinos take great precautions to protect themselves, such as building homes with walls made of heavy stone.

Monsoons

Monsoons are the seasonal periods of wind that bring rain to parts of the Philippines (and all of Southeast Asia). Because the rains are brought in by wind, it is normal for half an island or even mountain to get rain while the other half (on the rain shadow, or downwind, side) stays dry. There are two different monsoon seasons that blow across the Philippines. The northeast monsoon (called *amihan*) occurs from November to April, and the southwest monsoon (called *habagat*) comes from May to October.

3

The Philippines Through Time

People first arrived in the Philippines thousands of years before writing was invented and long before there were cameras or videos to document their lives. How do we know when people first arrived in the Philippines and what their lives were like? Finding out takes a little sleuthing. Archaeologists (people who study ancient artifacts) have found human bones and stone tools in caves and in the ground. From these clues, they can figure out who these people were, when the tools were made, what they were used for, and what life may have been like for these early peoples.

EARLY HISTORY

The oldest clues we have from the Philippines are some human bones and stone tools, which probably were used for hunting. These are about 50,000 years old, so we know that this is the earliest definite date that humans were there. Very little is known about these early people except that they probably walked to the Philippines between about 110,000 and 50,000 years ago. Walked? Hopefully you're thinking, "How could they have walked over the ocean to a bunch of islands?!" The country is a group of islands today, but they weren't necessarily islands thousands of years ago. Sometimes the ocean level drops, exposing bridges of land between areas that now are surrounded by water. For example, the ocean level dropped many times during the last ice age, which ended about 10,000 years ago. During that time period, most of the Philippines were isolated by water. One long skinny island called "Palawan," however, was connected by land to a very large island farther west, called "Borneo," which was connected to the Asian mainland. As you may have guessed already, the oldest human artifacts in the Philippines were found on Palawan. Also, the stone tools found with them have a strong resemblance to stone tools found on Borneo.

The Aeta Era

We know little about the first people to come to the Philippines, but we know a bit more about the Aetas. These people arrived about 25,000 years ago, probably also by walking to the Philippines from the Asian mainland. At this time, they already had spread throughout other parts of Southeast Asia. The main reason we know more about Aetas than about the earlier settlers is that Aetas still live in the Philippines, as well as in Sri Lanka and Malaysia.

Today, Aetas are commonly considered the aboriginal (or original) people of the Philippines. They are typically small in stature and dark-skinned and have curly black hair, all of which makes them look very different from later inhabitants of the Philippines. For this reason, Aetas are sometimes referred to as "Negritos," a term meaning "small, dark people" that the Spanish gave them. Aetas, however, most often refer to themselves as *kulot* meaning "curly" in reference to their curly hair. They once lived throughout the Philippines, but now they are found only in small communities, primarily on Luzon Island. Originally, the Aetas mostly lived on the coasts and subsisted on fish and other marine resources, but later immigrants who also wanted to live on the coast forced them to retreat inland. These once coastal people now live almost exclusively in and near forests. Today, they are some of the people who are most knowledgeable about the amazing natural forests in the Philippines.

Malay Immigration

People of Malay descent constitute the major portion of today's Philippine people. The Malays' origins are complicated and poorly understood, as is their migration to the Philippines. Malay immigrants are thought to have come to the Philippine islands during the period from about 2500 to 500 B.C. from the Asian mainland. These immigrants, sometimes referred to as Proto-Malays or Austronesians, arrived in long canoes from places such as Taiwan. They established villages around coastal areas and brought with them the first copper and bronze articles that were found in the Philippines. Although little is known about this time period, these groups definitely were skilled farmers, given their thousand-year-old rice terraces in the mountains of Banaue. The Malays went on to colonize much of Southeast Asia, primarily by boat.

Waves are a good metaphor for describing the demographic and cultural changes that occurred in the Philippines through its early history. Immigrants returning to the Philippines from about 500 B.C. to A.D. 1500 are a good example. These descendants of early Proto-Malay immigrants who later moved on to other countries doubled back and returned to the Philippines. They brought with them new cultural practices from contact with civilizations farther west. These new people and cultural practices had a profound impact on the Philippines that continues to the present day. They brought with them new forms of agriculture that used plowing, which are particularly suited to flat, lowland areas. In fact, the primary draft animal found throughout the country today, the water buffalo (known locally as the *carabao*), was brought to the Philippines on Malay boats hundreds of years ago. With help of carabao, early Malays cleared thousands of acres of lowland forests, turning them into fields of annual food crops. These stable food supplies, in turn, helped the Malays increase their population. Today, most of the lowlands in the Philippines are heavily populated by people of primarily Malayan descent.

Other Asian Influences

Through the first 1,000 years A.D., the Philippines was heavily influenced (often indirectly) by other cultures of people found in places as far away as India, China, and the Middle East. It was from India and China, for example, that the Malay people originally got their water buffalo, as well as other farming practices such as irrigation and rice cultivation. Buddhist-Hindu culture was integrated into the Philippines from A.D. 800 to A.D. 1478, as people from two Indian settlements in Indonesia immigrated into the country. There is not much evidence of this culture in the Philippines today, but there are many towns and areas named with Indian words.

Chinese merchants in the Sung Dynasty (A.D. 960–1279) came to the Philippines from Indochina to trade Chinese crafts such as porcelain for native wood and gold. It was from the Middle East that much of the region, including present-day Malaysia, Indonesia, and especially the southern part of the Philippines, became exposed to Islam. In 1380, an Arab scholar named Makdum arrived in the Sulu Islands, in the southern Philippines. His mission was to bring his "Propagation of Islam" to the Philippine islands, but he taught mostly on these southern islands. The movement was successful and, by 1475, a powerful Islamic center was established by Sharif Mohammed Kabungsuwan. This Muslim leader ended up marrying an influential native princess and became the first sultan of Mindanao. These areas of the Philippines are still the center of the country's Muslim faith.

SPANISH COLONIZATION

When Ferdinand Magellan arrived near Cebu Island in April of 1521, he had no idea that this would be his last stop as a great Portuguese explorer. He was the first European ever to gaze on the Philippines, and his discovery would change the country's history, as well as his own. His main interest was to find a trade route to the Molucca Islands (still far to the west) to acquire spices that at the time were all the rage in Europe. Second, he was spreading Catholicism and increasing the territory ruled by his employer, Spain. On landing, he put a wooden cross in the ground and began converting the people living on the islands from their native religions to Catholicism. He also declared that all these islands would now belong to Spain. The first couple of islands were fairly receptive to Magellan's plan, perhaps seeing his giant ships and many armed soldiers. These people allowed themselves to be baptized and converted by the strange white man who had arrived by sea.

The ruler of the largest village in the area (located at present-day Cebu City), Humabon, also converted. When his rival, Chief Lapu-Lapu of the neighboring island of Mactan, resisted converting, Humabon convinced Magellan that Lapu-Lapu should be attacked. Magellan agreed, still feeling the taste of success from having a whole island adopt the Catholic faith. He headed to Lapu-Lapu's home on Mactan Island, hoping to find a docile and small community that would be conquered easily. Instead, Lapu-Lapu had heard of

In 1521, Ferdinand Magellan was the first European to find the Philippines, and his discovery led to 300 years of Spanish rule in the country. Spain unified all the islands under one name and one government, and spread Catholicism throughout the country.

the attack ahead of time and rallied 1,500 men to face and defeat this intruder who had invaded their land and imposed a new religion. When Magellan and his crew arrived, they were in for a fight for their lives. Many of them, including Magellan, died in the battle. When Magellan's ship finally returned to Spain, more than a year later, the sailors had completed the first-ever around-the-world ocean voyage. Partly because of the encounter with Chief Lapu-Lapu, only 18 of the Magellan's original 270 sailors had survived the epic journey.

The Philippines Becomes a Spanish Colony

Twenty-some years later, in 1543, a second explorer working for Spain arrived in the islands. Roy Lopez de Villalobos declared the entire archipelago a conquest and in doing so named the islands "Filipinas," after King Philip II of Spain. Until this time, the islands had operated separately from one another. Each island or set of islands had its own rulers and had relatively little to do with other islands, so, although the Filipinas islands resisted this colonization on an island-by-island basis, they did not have the organization needed to stand against their invaders collectively. By 1565, permanent Spanish occupation of the Philippines began. Spanish fleets coming from Mexico, another Spanish colony, began to arrive regularly. In addition to conquering islands one by one, the Spanish built forts and churches and spread Catholicism throughout the country.

The 300 years of Spanish rule had a huge impact in the Philippines. For the first time, this group of more than 7,000 islands was gathered together under one name, with one governing body. The Spanish chose Manila, with its big protected bay and large expanse of agriculturally productive land, as their capital city. In general, agricultural plantations got their start during the Spanish era, because foreign investors were allowed into the country for the first time. These foreigners amassed huge landholdings and generated agricultural products such as sugar, which were sold to European customers. Manila served as the main trading post between Spanish colonies in Mexico and the Eurasian continent, with goods, especially those from China, in high demand. Manila remains the capital and the main commercial hub of the Philippines to this day.

The Spanish unified this archipelago with religion. Through the conversion of people and the establishment of churches, suddenly Filipinos had much more in common. Spanish friars in the Philippines exerted enormous influence over local populations, controlling not only religion but also tax collection, public-works projects, and law enforcement. As part of their cultural conquest, the Spanish gave Filipinos Spanish surnames. Island by island, they assigned recognizable Spanish surnames to each and every family. Because most of these people had not previously used surnames, the last names of Filipinos today are still largely Spanish.

Finally, the Spanish occupation inadvertently ended up uniting the Filipinos from their separate islands by giving them a common enemy. Struggling against foreign rule was enough common ground for Filipinos to bind together for the first time as a nation, opposing the Spaniards' corruption and tyranny. How to deal with the Spanish was the question. The Philippines' most famous hero, José Rizal, argued that reform was possible while the Spanish remained in power. Nonetheless, when those who believed that rebellion was the only way to independence revolted, the Spanish arrested Rizal. In 1896, they executed him by firing squad in Manila's Luneta Park.

THE UNITED STATES' SHORT RULE

In 1898, the United States won a war against Spain, fought primarily over Spain's occupation of Cuba. With the victory, the Philippines was transferred from Spanish to American control. At the end of the Spanish-American War, despite significant Filipino effort to take up arms and fight the Spanish for their independence, the Americans made a surprising decision: Their victory against Spain meant that they had won the right to colonize the Philippines.

Unfortunately for Filipinos, this meant that all their fighting for independence against the Spanish was in vain; they essentially had traded one imperialistic power for another. Nonetheless, they turned their fight toward the Americans. Led by General Aguinaldo, the Filipinos fought tenaciously against superior forces. By 1901, Aguinaldo was captured and 16,000 Filipino soldiers and more than 200,000 Filipino civilians had been killed. The Filipino-American war, the United States' first overseas war, was over.

The United States was a newly developing Western power and was enthusiastic to have a colony, much like other Western nations. Although like their European peers in their zest for imperialism, the Americans had a much different approach to colonization. They immediately set out to improve their new colony, bringing in American teachers to teach English, establishing democratic institutions, and undertaking construction efforts. They built new schools, roads, bridges, and an electrical infrastructure throughout the islands. Because of the many benefits they brought, the American presence generally is remembered well by Filipinos. Of even greater importance is that U.S. soldiers fought side by side with Filipinos to root out the Japanese during World War II.

IMPACT OF WORLD WAR II

In 1941, shortly after they attacked Pearl Harbor, the Japanese attacked Manila. This time, instead of fighting against their colonizers, Filipinos joined forces with the Americans in battle. Both countries lost many men and eventually the Philippines succumbed to Japanese military rule. They suffered under the brutal Japanese power. A famous example of this treatment was the Bataan Death March in which the Japanese forced 80,000 Philippine and American prisoners of war to march to a prison camp 65 miles (105 kilometers) north of Bataan Province. During the march, 10,000 men died due

With its victory in the Spanish-American War in 1898, the United States acquired the Philippines from Spain. The United States then fought Filipino soldiers to secure the country as its first colony. Here, American soldiers stand on a rampart in Manila in 1898.

to starvation, disease, and exhaustion. It was not until 1944 that the tide turned, with the promised return of American General Douglas MacArthur. Four years earlier, MacArthur was forced to flee the country, but he made his famous promise: "I shall return." In 1944, when MacArthur came wading ashore on the Philippine island of Leyte to liberate the country from the Japanese, he secured his place as a special hero to the Philippine people. Overall, the war created a relationship between America and the Philippines that was unique in their shared history. Shortly after the end of the

The Philippines were an American colony until the Japanese military took control early in World War II. In 1944, American General Douglas MacArthur, seen here, center, came ashore at Leyte to recapture the Philippines and liberate the country from the Japanese. After the war, in 1946, the United States granted the Philippines independence.

war, on July 4, 1946, the United States granted the Philippines full independence. (For additional information on Japan and the Philippines during World War II, enter "Philippines Japanese occupation" in any search engine and browse the sites listed.)

The Philippines has emerged from its rich and complicated history as a vibrant, unified, yet culturally diverse country that is one of the most unique in all of Southeast Asia. Considering the extent to which the Philippines has been influenced by other nations and the great diversity of its citizens' origins, it is surprising how much of the country's culture is distinctively Filipino. How the Philippines continues to remain unified and yet celebrate democracy, diversity, and global influences will be a fascinating part of its history to come!

CHAPTER

4

People and Culture

T he country of the Philippines is a collection of islands, each with its own people and communities. The influences of Spanish and American colonization did a lot to homogenize (make the same or similar) these different cultures. Struggles against these powers, as well as against the Marcos regime, unified the islands and brought forth the first real signs of nationalism. These influences and the long history of the Filipino people have created a diverse blend of cultures with certain common themes. In this chapter, you will learn about the people of the Philippines and their national culture, as well as the ethnic cultures that are still present. Generalizations about Filipino society as a whole will be made, and the different ethnic, economic, and religious groups that make up that society will be described.

SOCIETY

An estimated 85 million people live in the Philippines. It is amazing to realize that, although the country is about the size of the state of New Mexico, the Philippines has nearly 50 times New Mexico's population! Population density is a good measure of how crowded a country is, because it is the population size divided by the amount of space that population lives in. With a population density of about 725 per square mile (255 per square kilometer), the Philippines is one of the more crowded countries in the world.

Not only does the Philippines have a large population for its size, but its population is growing very rapidly. Population growth is measured by the number of new people added to the population each year minus the number who have left or died. The population growth rate in the Philippines is currently about 2 percent, which is also one of the higher growth rates in the world. This is over twice the United States' growth rate (.9 percent). There are two main reasons Filipinos give for having many children. Catholic beliefs suggest that birth control is against the will of God, and most Filipinos are Catholic. Also, like people in many Asian countries, Filipinos want as many children as possible to secure their future well-being. More children means more helpers in the fields and more protection for parents in their old age.

The age structure of Filipinos is skewed toward the younger generations. More than half of the 85 million Filipinos are under the age of 20, and less than 7 percent of the population is over 65 years old. In the United States, only about a quarter of the population is under 20 years old, and 12 percent is over 65. Because overpopulation is becoming a concern in the Philippines, age structure of the current population is very important. Even if effective family planning programs are implemented, there will be a long lag time until the population levels off as this huge younger generation matures. It is a good thing that Filipinos seem so tolerant of crowded conditions, because they are likely to be even more crowded in the future.

SETTLEMENT DISTRIBUTION AND PATTERNS
Urban Life

The Philippines' 85 million people are distributed unevenly in the country. Forty percent of all Filipinos live in cities. The capital city, Manila, has an estimated population of about 18 million people, nearly one-fifth of the country's people. Urban life in the Philippines has many qualities similar to living in cities anywhere. People are packed into a fairly small area and live most often in big concrete apartment buildings. They have a variety of jobs: taxi driving, entertaining, working in offices, peanut selling, and everything in between. Most of the workers in Manila lived somewhere in the provinces before moving to the crowded city, where they could make more money.

Provincial Life

Many Filipinos live in the much less crowded countryside, called *probincia* (provinces). People in the provinces tend to earn a living more directly from the land. They have livelihoods of fishing, farming, and forestry and have individual homes built out of local products like wood, bamboo, and thatch (strong leaves). Most thatch comes from a plant called *nipa,* which is a palm tree that grows on the coasts in saltwater marshes. Nearby there often are larger towns where farmers go to do their shopping and banking and where they can sell their products.

On the coasts of the islands, most fishermen fish from their own *bangkas,* long, wooden, canoe-shaped boats with bamboo stabilizers on the sides. They head out to sea at night with their outboard motors purring, shine bright lights on the water, and wait for fish to get tangled in their long fishing nets. Then, in the early dawn, they gather the nets and bring the fish back to shore, selling them at the bustling morning markets.

Most of the agricultural land in the Philippines is in the flat fertile river valleys. These areas have rich soils, which promote easy farming of vegetables and fruit trees. They also have access

Forty percent of Filipinos live in cities, but the rest are scattered through-out the countryside and earn a living off the land, like these fishermen bringing their boats back to the shore of their remote fishing village after a day of work.

to irrigation from the river, which is crucial for rice growing. Most family farms have a number of different crops growing to cover many of the food needs of the family, in addition to some chickens and pigs raised for eggs and meat. Often, a farmer will choose a couple of crops to grow in bulk as cash crops. He or she also may have some extra pigs being fattened for later sale. When these animals mature and the crops ripen, they are taken to market to be sold.

The smallest numbers of Filipinos live high in the mountains. Communities in these regions are difficult to reach and therefore are very remote from the rest of the country. People in these regions are often of tribal descent, meaning that they come from different ethnic backgrounds than people of the lowland communities. Probably because they are so isolated, mountain communities are very dependent on local farming for much of their food. These communities use terraces to create flat fields on their hillsides to plant their rice and vegetables. Like lowlanders, they also have chickens and pigs for eggs and meat. Villages tend to be located along rivers, which they use for irrigation, to clean their laundry, and to hold small fish pens for freshwater fish like tilapia.

Settlement Trends

Because of dwindling resources, farmers today are venturing farther than ever before to acquire good land for farming. Forested areas, especially near towns and along roads, are being converted to farmland through slash-and-burn techniques, as well as through large-scale logging. With the ever-mounting needs of a growing population and the little control by government, soon very little natural forest will be left.

Many of the Philippines' 85 million people live in the provinces, but this is not a static system. The past several decades have witnessed a constant flow of Filipinos from the country to the cities. This is a phenomenon all over the world and is caused by a number of factors. Many Filipinos are moving to the cities for the job opportunities and higher wages that urban areas offer. Also, many young Filipinos are not interested in working in the fields like their parents. They move to the cities in the hope of putting their education to use in careers in areas such as computer science and information technology.

A third trend in the Philippines' population shift is the outflow of workers to overseas jobs. Each year, hundreds of thousands of people leave the Philippines with work visas to try to earn larger wages in other countries. In fact, about 1 in 10 Filipinos works *outside* the country. They seek jobs like nursing, house cleaning, and construction and commonly work in countries such as Saudi Arabia, Australia, and Canada. In addition, many Filipinos leave the country for good, taking up residence elsewhere. Filipinos are the second-largest Asian immigrant group in the United States today, behind the Chinese.

ETHNICITY

Most Filipinos are descendants of the Malay people who migrated to the country several thousand years ago. About 12 percent of the people in the Philippines make up the cultural minority groups, called tribal Filipinos. There are 60 recognized ethnic groups, including the aboriginal Aetas (Negritos), the many mountain tribes in Luzon, and the Muslim communities in Mindanao.

Aetas once lived throughout the Philippine islands. As the high volume of new immigrants arrived in the country, however, they were either integrated into lowland society or, more often, pushed back into the forests. Now, true Aeta communities can be found most often in the country's remote inland areas, where they still practice their traditions of hunting and gathering from the forest.

The many mountain tribes of Luzon are largely of Indochinese descent and live very different lives than the Malay Filipinos. Most of the tribes are descendants of infamous head-hunters, known to seek revenge by chopping off the heads of their enemies, although head hunting is not common anymore.

The Ifugao tribe of the Central Cordillera (mountains) has developed an impressive farming practice, rightfully said to be the "eighth wonder of the world." By hand, they built terraces of irrigated fields for rice growing. These line entire

The Ifugao tribe, one of many mountain tribes in Luzon, has an impressive farming practice that uses terraces of irrigated fields to grow rice. Members of the tribe, in ceremonial clothes, stand overlooking the terraced fields lining a valley in Banaue, Ifugao Province, North Luzon.

mountainsides like staircases for giants. In the tribal areas near Bontoc are another well-known tribe, the Igorots. When an Igorot tribe member dies, he or she is put on display on a chair, known as the death chair, to allow other tribe members time to pay their respects. Finally, days, even weeks, after the death, there is a ceremony in which the chair with the dead person is raised into the air and carried for miles to a burial site. Young tribal men take turns carrying the heavy chair as far as they can, in the traditional belief that by doing so they absorb the strength of the person they carry. (For additional information on the rice terraces, enter "rice terraces" into any search engine and browse the sites listed.)

RELIGION

From the accounts of Spanish explorers in the 1500s, the religion of early Filipinos was a sort of animism (a belief system wherein natural things like animals, trees, and stones are inhabited by spirits). Although Catholicism has officially replaced animism, there are still some signs of animism evident in contemporary Filipino culture. Many Aeta communities hold animist beliefs, and they usually are quite willing to point out the special powers of nature. It is also common for non-Aeta Filipinos, especially in the provinces, to talk of spirits, dwarves, and witches of the forest. Some of these mythical creatures serve to explain particular phenomena. For example, if someone gets lost in the woods, he or she may blame it on a *kapre*. This is a gigantic human-like male creature, about 10 to 12 feet in height and with a body and face covered with hair, who supposedly lives in towering forest fig trees and smokes a cigar.

Today, the most widely practiced religion in the country is Catholicism, with more than 83 percent of Filipinos belonging to that religion. Many of the national holidays in the Philippines are Catholic holidays. Every Sunday, throughout the country, you can see people headed to church with family and friends.

About 5 percent of Filipinos are Muslim, and most of them live in the southern Mindanao and Sulu areas. People of these communities are descendants of the Arab Muslims who arrived in the area in the late fourteenth century and established Muslim spiritual centers. This area of the Philippines remains a stronghold for Muslims, and, because of the stark contrast in their beliefs, many Muslims would like to separate from the rest of the Philippines. They believe that the Catholic Filipino government does not recognize their special needs. Some Muslim groups resort to violent methods of protesting the Filipino government, using bombs, kidnapping, and other scare tactics, many of which make the international news in the United States.

Other Christian religions, such as the Church of Christ, Philippine Independent Church, Baptists, Mormons, and Jehovah's Witnesses, make up another 9 percent of the population. Finally, among the Chinese Filipinos, there are some practicing Buddhists, most of whom are found in Manila.

LANGUAGE

You can imagine that language could be difficult in a country of thousands of islands and so many cultures. The different immigrants who came and settled in the Philippines brought their own languages with them. In fact, 169 different languages and dialects are spoken in the country. A dialect is a form of a language spoken differently and using different words. The kinds of English spoken in Scotland, England, and Australia are all dialects different from that in the United States. If you count just the dialects alone, there are more than 80 spoken by people in different regions of the Philippines.

Even though the Spanish lived in the Philippines for about 400 years, little Spanish is spoken there today. During the Spanish occupation, Spanish was actively taught to only Filipino students in the elite social classes. By 1968, Spanish was no longer a required course in high school and college, and today there is very little pure Spanish heard in the Philippines. There are some Spanish words in most Filipino dialects, especially for money, numbers, the time of day, kitchen utensils, furniture, and foods.

It was not until 1946 that a national language was designated for the Philippines. There were so many different languages and dialects that it was difficult to choose one to serve the whole country. Finally, Filipino (based on a local language called Tagalog) was chosen as the official national language. It probably was selected because it is the main language spoken in the capital city of Manila. Many Filipinos resent this choice, feeling that their local language should have been chosen. Cebuano (also known as Visayan) is a language spoken on many of the islands in the center of the country. The people in this area have argued that more Filipinos speak Cebuano than Tagalog, so this should be the national language.

In some ways English is the unofficial national language of the Philippines. With so many different Philippine languages, the language of the American colonists served as a noncontroversial alternative to any one particular Philippine language. Still, the president's speeches are usually in Filipino (also called Pilipino), as is the news. Some national newspapers are written in Filipino, and most radio and televised news programs are aired in Filipino. Most entertainment meant for the whole country, including movies and radio and television programs, uses Filipino.

Even though there now is a national language, there is little chance that other Filipino languages will disappear. When people need to communicate across regions, they are able to converse in Filipino or even English. When at home in their local community, however, people prefer to speak in their local dialect or language. This is their mother tongue and the language associated with all of the cultural traditions of their home region. Because of this practice, most Filipinos know at least two languages, and often three or more.

5

The Philippine Government

HISTORICAL GOVERNMENTS

Before the arrival of Spanish in the 1500s, the Philippine islands were independent rather than a unified country. How these islands were governed is poorly known, and what we do know is an interpretation by Spanish observers. Most of these islands' communities seemed to have a similar type of social and political organization. The basic unit of a community was a kinship (related) group of people called a *barangay*. The word "barangay" comes from the Malay word *balanghai*, which is a large ocean outrigger boat that could hold up to 90 people. The Malayan immigrants, who traveled together on these balanghai, also settled in communities in the Philippines together, and therefore, they called their neighborhoods barangays. At the head of the barangay was the *datu* (chief), who was an elder in the community. The barangay lived as a cooperative group, farming the same lands, helping each other, and sharing the same benefits. There was no sense of individual ownership, because everything they needed belonged to the barangay as a whole.

Spanish Colonial Government

Spanish colonization brought major changes to the government of the Philippines. Many of these still play influential roles in how the Philippines' government works today. Under Spanish rule, the Philippine islands were united under one government for the first time. The Spanish government oriented the country toward a singular governing body based in Manila. Filipinos got used to having the main government located in one spot, so even when the Spanish left, Manila remained the capital.

Second, religion and politics (more exactly, the Spanish Catholic Church and Spanish colonial government) were completely linked in the Philippines. Many of the rulers went through the Catholic bishops to carry out policy. The Catholic bishops, in turn, became corrupt with so much power. This combination had a profound impact on Filipinos; today, the Catholic Church still plays very powerful roles in the Philippine government.

Third, the Spanish introduced the idea of land ownership to replace the previous system of communal ownership. On taking control of the Philippines, the Spanish immediately proclaimed that all Filipino land effectively belonged to Spain. The Filipinos who grew food on those lands for subsistence became Spanish servants.

Finally, to rule all the different communities on the different islands, the Spanish made use of the old governing system of datus. They gave new power to the datus so they would act as their liaisons (contact people) in the provinces. They transferred the ownership of the communal land to the datus' families, so they would watch over these lands for Spain. Because the Spanish gave the datus so much more power than their fellow community members, they created a Filipino upper class that had never existed before the families of these datus were suddenly part of a privileged group that had all the land and therefore a lot of control over the Filipinos in their area. Because these families were put in charge of their entire communities, when the Spanish left, they took over. Today, the Philippines is run mostly by these few prominent families, who have most of the wealth and power in the country.

THE BEGINNINGS OF NATIONALISM AND DEMOCRACY

By the end of the nineteenth century, Spanish tyranny had become so corrupt that many Filipinos spoke out to bring people together in opposition to this foreign rule. Many of these outspoken "nationalists" later became heroes of a revolution against Spain. Even Chief Lapu-Lapu, who had been forgotten for some time after his battle against Magellan, became an honored hero because of his success in stopping the invasion of an imperialistic attempt.

Eventually, the Philippine nationalistic movement gained so much momentum that a revolution against Spain was inevitable. In late 1897, the Filipinos began their struggle for independence. In April of 1898, when the Americans began to fight the Spanish over an incident in Cuba, the Filipinos took advantage of this disagreement. They hastily jumped on the side of the Americans to fight and eventually defeat this common enemy. On June 12, 1898, General Aguinaldo, a Philippine nationalist who led the Philippine troops, declared the Philippines independent.

When the Americans took control of the Philippines in 1898, they did not expect to stay long. In fact, they hoped that their administration would be temporary, lasting only until the Philippines seemed ready to govern itself. They did many things to encourage the establishment of a government that was both free and democratic. In support of this future government, the American officials focused on promoting public education and the development of a reliable legal system. As early as 1907, the first legislative assembly was elected, made up of all Filipinos and largely left under Philippine control. The United States also promoted the formation of a civil service run by Filipinos.

On July 4, 1946, Filipinos raised their flag and lowered the American flag as a symbol of their newly independent nation. Though it was no longer under American rule, the Philippines was still greatly influenced by the United States in everything from the development of its government to its language.

World War II was very destructive in the Philippines, leaving more than a million Filipinos dead, Manila bombed out, and the country generally shaken. Both the United States and the Philippines wanted to go ahead with their plans for Philippine independence, however. On July 4, 1946, the Republic of the Philippines became an independent country.

THE PHILIPPINES' CONSTITUTIONAL DEMOCRACY

Stemming from much American encouragement, the Philippines' government has many similarities to government in the United States. It is democratic and has a constitution that is one of the longest in the world. As is true of government in the United States, there are three branches in the Philippine government. The head of the executive branch is called the president and is seconded by the vice president. The legislative branch, also called Congress, has both a Senate, with 24 elected officials, and a House of Representatives, with a total of 250 members. The highest ruling court in the judicial branch is the Supreme Court, which is made up of justices appointed by the president.

Under the national level, the Republic of the Philippines's administration is divided into 12 regions, consisting of 73 provinces. Like U.S. states, every province has a capital city where the provincial government, with the governor at the head, conducts business for the province. Each city within a province has a mayor, and within that city are several barangays. At the head of each neighborhood in a Filipino city is the barangay captain, who is responsible for representing the neighborhood's interests to the mayor.

CURRENT GOVERNMENT

To understand the current government in the Philippines, it is important to understand the very recent history of the Philippine government. Most people who are even slightly familiar with the Philippines are well aware of the history with Marcos.

A Corrupt Dictator

Ferdinand E. Marcos was elected president in 1965 and then reelected in 1969. Although the Philippines was finally on its own, this was a time of disorganization, corruption, and instability. The Marcos administration brought some order to the chaotic country, but when Marcos was supposed to end

Ferdinand Marcos, seen here at his 1965 inauguration as the sixth president of the Philippines, brought some order to a country that had been plagued by instability and corruption since gaining independence. However, in 1972, at the end of his two-term presidency, he refused to give up power and ran the country as dictator until 1986.

his term as president in 1972, he refused to leave by declaring martial law over the country. Martial law amounted to Marcos using the military to remain in power so he could implement his idea of "New Society." His troops used weapons to impose curfews at night. They took all guns and weapons from Filipino citizens and controlled what people learned in the media by allowing only censored news articles to be published. At first, this appeared to be an effective strategy. Crime rates went down, public health increased, and the Philippines joined many important international organizations. After a short time, though, it became clear that Marcos and his wife were addicted to the power that comes with living in the Malacañang Palace (the president's home in Manila). They lavished themselves with riches, gave political positions to their friends, and lived the life of royalty rather than that of publicly elected politicians.

People Power

With all of the restrictions and the obvious corruption, many Filipinos began to question their leader. Several groups used violence to voice their opinions, and many opposition parties sprung up. One of these political rivals was a man by the name of Benigno Aquino, Jr. "Ninoy," as Aquino was nicknamed, was a popular politician who opposed Marcos' dictatorship. Threatened, Marcos had Aquino imprisoned for seven years. Then, in 1980, Aquino received permission to go to the United States for needed medical treatment; Marcos hoped he would remain there and be less of threat to his rule. In 1983, Benigno ignored warnings to stay in exile and flew back to his homeland. As he stepped from the plane on arrival in the Philippines, he was shot from behind by government soldiers.

Enraged by the assassination, many Filipinos quickly threw their support behind Aquino's wife, Corazon, who took up where her husband had left off in criticizing Marcos' rule. In 1986, she ran against Marcos in an election in which Marcos' corruption was sure to prevail. The night before the election, however, thousands of Filipinos poured into the streets of Manila for a nonviolent protest against Marcos' power. The protest lasted many hours and grew quite large and it included many defectors from Marcos' military. At the end of what was called the "People Power" movement, it was clear that Marcos would no longer dictate over the Philippines.

Corazon (Cory) Aquino became president in 1986 and restored democracy to the Philippines. Because of her bravery in surviving her husband's death and standing up to Marcos, she is a national hero and symbol of democracy and people power in the Philippines. After her presidency ended, she nominated Fidel Ramos as her successor. Ramos won the elections and served as president from 1992 until 1998. In 1998, Joseph Estrada won the presidential seat, running as the representative for the "common man." His corrupt practices and cronyism (giving jobs to friends), however, caught up to him quickly. Just two and a half years into his term, in December 2000, angry Filipinos organized another mass demonstration to protest his administration. This "People Power II" march was much smaller than the famous march that overthrew Marcos 15 years before. It was effective in achieving its goal of ousting the president, however, and Estrada was jailed for embezzling public funds. The vice president and leader of Estrada's opposition, Gloria Macapagal Arroyo, quickly assumed the office of president after the protest, taking over in the middle of Estrada's presidential term.

SUCCESSES AND FAILURES OF GOVERNMENT

Just as Americans elect political administrators, the Philippines also holds regular elections to democratically choose its leaders. Because the president can lead the country only through one term (six years), presidential elections are held every six years. Many of the provincial leadership positions are also elected at this time. These elections tend to be much livelier than elections in the United States. For one, there are usually many more candidates running for the presidential and vice-presidential slots. Instead of just 2 likely candidates for president, as in the United States, the Philippines tends to have 5 to 11 candidates who could very well become president. Also, unlike the United States' combined presidential and vice-presidential tickets, these positions are elected separately in the Philippines. This means that the Philippines' elected president and vice president are often not from the same political party.

A high rate of voting citizens is a good sign that people believe in the voting process and in the results of the elections. Every country with political elections is interested in how many of their people had a say in the outcome through voting. Typically, about 70 percent of eligible Filipinos vote in elections. This is one of the highest voter turnout rates in the world.

Despite the large local participation in Philippine elections, there are some negatives to this process of choosing leaders. Often, violence and corruption surround elections. Large numbers of political candidates are assassinated each election year. There are also clear cases of corrupt campaign practices that lead to unfair election results. Despite efforts to watch over the tabulation of votes, claims of cheating are common at most levels. Local politicians have been known to buy votes by paying each person who votes for him or her, and there are many mysterious cases of candidates dropping out of the election for no reason.

A constant complaint against the current Filipino government comes from the Muslim populations in the south. More than one-eighth of the Filipino population is Muslim (more than 12 million people), so it is clear that some of the government's focus should be on their needs. Because the Muslims live far from Manila and because the government is dominated by the Catholic majority, however, the Philippine government does not focus much effort on the Muslim areas. To make matters worse, some Muslim insurgent factions of the southern population resort to violence to promote separatism. Because of the bombings and kidnappings they have used to demonstrate their intents, many Filipinos, including much of the government, are afraid of the Muslim region of the Philippines.

Another shortcoming of the Philippine government is its high level of corruption. Theft, deception, cronyism, nepotism (giving money and power to family members), and other types of corruption are commonplace. Such conditions make it difficult for the Philippines to move ahead and keep the Philippine government from being effective, because funding and energy set aside for important public services like education, postal communications, and public safety are often lost to corrupt officials.

Corruption also makes it difficult for other countries to help the Philippines. They do not want to invest in the Philippines if their aid will be lost to corruption. Annually, each of the world's countries is given a score as a measurement of its corruption to help foreign investors and aid agencies decide where to put their money. For several years, the

Philippines has rated as one of the most corrupt countries in the world. This means that foreign aid and investment is kept low, which has a negative impact on the Philippines' economy.

KEY POLITICAL FIGURES

Since the time of local barangay politics, Filipinas (female Filipinos) have played large roles in the Philippine government. During recent elections a higher percentage of registered women voted than men. There are women on the ballot each election year, and two of the last four Philippine presidents were women (as of 2004). About 18 percent of the congressional representatives are women, and about the same percentage of women hold cabinet positions in the national government. Although there are fewer females than the males in the top-level positions in the government, more than half of government bureaucratic positions are filled by women. At second-level positions in the government, nearly three-fourths are women.

In many ways, the political system in the Philippines is an oligarchy (government controlled by a few wealthy families). About 10 Filipino families hold more than 50 percent of the country's wealth. All of these families are descendants of the people given power by the Spanish. Because it takes power and money to run for office, it is not surprising that well over three-fourths of the high officials in government come from families among the Philippine elite.

The Spanish government operated hand in hand with the Catholic Church, and the Church still plays a strong role in the Philippine government. Cardinal Jaime Sin was instrumental in igniting the People Power movement, which overthrew the Marcos regime. As the top Catholic figure in the Philippines, he is influential in the presidential elections, because millions of Catholics vote according to his endorsements.

CHAPTER

6

The Philippine Economy

A country's economy can be understood in a number of different ways. In terms of international importance, we can look at what role the country plays in the global economy. A country's imported and exported goods, as well as its reliance on foreign investment and aid, are good measures of how the country operates globally. In terms of a nation's internal economy, we can investigate the different sectors of the economy, including natural resources, manufacturing, and services.

Another insight into a country's economy is how it affects the everyday lives of its people. In the United States, the role the country plays in the global economy seems fairly remote from our personal lives. Our reliance on resources, industry, and services may be a little closer to home depending on where we live and what our families do for work. When we think of how the economy directly affects us, though, what comes to mind are things like how much money we make, how much things cost, and who has (or does not have) much of the money in the United States.

In this chapter you will look at the Philippine economy describing its role in the global economy, as well as the internal structure of its national economy. You will get an idea of how the economy affects the average Filipino. Minimum wages, cost of living, and buying power, as well as a skewed distribution of wealth, all affect the daily lives of people in the Philippines.

ROLE IN THE GLOBAL ECONOMY

Other than Singapore, Hong Kong, and Brunei, much of Southeast Asia is considered "poor" or "underdeveloped" from an economic perspective. In this manner, the Philippines is no exception. One measure of economic health is gross domestic product (GDP), which is the total value of all goods and services produced in a country in a year. The Philippines ranks twenty-fifth for its GDP. This is in the upper 15 percent of the world's countries and is higher than many countries that are considered wealthy. The Philippines has a very large population compared with these other countries, however, so when the per capita (per person) GDP is considered, its rank is 129, which is in the lower half of the world's countries. This means that Filipinos have far less income than people in most of the world's countries.

Economists use a measure of income called purchasing power parity (PPP) to compare the buying power of average individuals in different countries. The annual PPP for an average Filipino is the equivalent of $4,178. In comparison, the PPP of an average American is $35,056, or nine times the buying power of a Filipino! In fact, 40 percent of all Filipinos live below the poverty line.

Exports and Imports

Another way to look at the Philippines' role in the world's economy is by looking at its imports and exports. Generally, yearly Philippine exports and imports are nearly equal in value. Most of the Philippines' exports are sent to the United States and Japan, with some other countries in Asia importing the rest. Top exports include electronic equipment, machinery, chemicals, and garments. Many American brands, such as The Gap, American Eagle Outfitters, Old Navy, Calvin Klein, and Ralph Lauren sell clothing put together in the Philippines. The next time you put on a name brand T-shirt, check the tag to see if it was made there.

There are also many agricultural products exported from the fields of the Philippines. Coconut products, tobacco, Manila hemp, bananas, pineapples, and orchids are commonly exported to countries like the United States. Wicker and bamboo products like mats and baskets from the Philippines can also be found here. Keep your eyes at import stores like Pier 1, because you are bound to see many "Made in Philippines" stickers.

The major Philippine imports are raw materials, equipment, fuels, and chemicals, which come mainly from Japan and the United States. Rice is another product the Philippines imports in large amounts. Despite having more than 2 million rice farmers, the Philippines is no longer producing enough rice to feed its growing population. Its yearly total imports of rice currently exceed 1.2 million metric tons, which come from other Asian countries including India, Thailand, Vietnam, and Pakistan. As you can imagine, the Philippines does not want to be dependent on other countries for rice, which is a staple of the Filipino diet. Scientists in the Philippines are working hard to develop more productive varieties of rice, hoping to achieve self-sufficiency.

Foreign Assistance and Debt

A final aspect of the Philippines' economy that has global importance is its reliance on foreign funding. External debt has plagued the Philippines since the lavishly corrupt Marcos era. With an economy that leans on foreign input, the Philippines spends a lot of its money paying off interest from past loans rather than bettering the country. Dependence on foreign help also detracts from the Philippines' goal of self-sufficiency and

The Philippines mainly exports garments, machinery, and agricultural products like fruits and tobacco. Though the country has over two million rice farmers, like these men working in Nueva Ecija, its rice crop can no longer feed the population and rice actually has to be imported from other Asian countries.

independence as a country. Currently the Philippines has external debts that exceed $60 billion, which puts it in the top 19 countries in the world for owing money.

Foreign Investment

While the Philippines parts with a lot of its money to service its debt, money pours in from outside through foreign business investments. Historically, the United States has been a major source of foreign investment in the Philippines. Recently, this has declined in part because of Philippine laws against too much foreign ownership. Japan and Taiwan now invest the most in the Philippines, with the United States, Germany, and China as distant runners-up.

LABOR FORCE

Despite being a rather small country in size, the Philippine labor force ranks fourteenth in the world, at about 34 million. The rate of unemployment, however, is currently at 10.2 percent. Although this figure is relatively high, it is not as alarming as it sounds. Many unemployed Filipinos work informally, selling products at the market or doing other services, and are not counted in labor statistics. Another thing to consider is the Filipino work ethic, which is different than we are used to in the United States. In the Philippines, it is not expected that every member of a family work. Families usually are cooperative, allowing different members to play different roles. Some members may stay home to take care of children and cook and clean, and others may take classes. Most families rely on the income of just a few working individuals to keep the family going. Therefore, although the number of working age Filipinos may be very high, it does not imply that all of these people aim to have full-time jobs all the time.

An important part of the Filipino work force is its overseas workers. These workers are an important source of national income because of the large sums of money they send back home. At any given time, several million Filipino citizens are outside the country working as nurses, engineers, domestic helpers, teachers, sailors, entertainers, and other roles. There were nearly 8 million Filipinos working in 162 different countries in 2002. Remittances (the money they send back to the Philippines) amounted to $6.2 billion that year, which was 8.2 percent of the country's gross national product.

LAND AND NATURAL RESOURCE USE

Countries differ in their economic strategies. Some of these differences are based on where the country is and what raw materials it has to work with. Each country has a store of economic wealth in the use of its land and natural resources. In the Philippines, two-thirds of the people live off the land by fishing, farming, and forestry.

Fishing

Filipinos make their living using the country's vast fish resources on every scale. There are fishermen who just fish for their family's meals and may bring a little extra back to sell at the local market. In the middle of the fishing industry are the Filipinos who have made jobs out of buying fish locally and distributing it to more expensive markets in areas such as Manila. At the top end of the fishing industry are the wealthy fish exporters. Some sell live fish to markets in China, and others process fish by canning it or drying it and then export their products throughout Asia.

Agriculture

The agricultural lifestyle in the Philippines is similar to the fishing industry in its scales of wealth. Most people in the Philippine provinces are farmers on a simple scale. Many provincial homes have a few fruit trees in their yards. These may include guava, mango, papaya, rambutan, and pomelo. Provincial houses are also likely to have some vegetables, such as squash, mungo beans, tomatoes, and eggplant, growing nearby. If a family has a little extra money, it might invest in chickens, goats, or pigs, which are grown to share at a future fiesta, wedding, or other celebration. In this way, whether or not the family derives an income from another source, it will always have food to rely on at home.

Farther up the farming ladder, there are other people making money from farming. In the mountain regions, most families plant many vegetables and rice so that some can be sold at the market for money. In the lowland areas, rice is also a cash crop, as are all sorts of fruit trees, melons, and a few types of vegetables. Finally, the wealthiest people in the farming industry are those who own packaging plants that mass-produce Philippine farm products for sale overseas. In the United States and all over the world, Philippine bananas, pineapples, mangoes, dried mango slices, and dried coconut can be found in grocery stores. Even more Philippine farm products (such as coconut oil and other palm oils) can be found as ingredients in foods we buy here. More than half of the world's coconuts are produced in the Philippines.

Forestry

For most of the 1900s, forestry was an important part of the Philippine economy. During the last century, nearly all of the valuable Philippine hardwood trees were cut and sold at top dollar to countries such as the United States and Japan. As a result, a handful of people became very wealthy from the logging industry and timber sales were a powerful force in the economy.

Now, since more than 95 percent of the Philippines' original forests have been cut, Filipinos have to resort to different ways to meet their needs for wood products. There are quite a few plantations where fast-growing trees are grown for wood. The wood produced on these farms, however, plays a minor role in the economy compared with the timber businesses of the 1900s. (For additional information on the destruction of Philippine forests, enter "Philippines deforestation" into any search engine and browse the sites listed.)

A more influential use of forests in the economy is the widespread harvesting of nontimber forest products. Communities located in the forest or on forest edges have made use of many of the plants that grow naturally in forest areas. Local gatherers collect things like rattan (for furniture), abaca (also called Manila hemp), bamboo, and nipa, which is used to weave walls and roofs for local houses. These products do not grow in large numbers as if they were farmed as a crop, but they can be plentiful in the wild and provide substantial livelihoods for those who collect them.

Minerals

Some of the Philippines' economy is also dependent on minerals found in the country. The most important of these minerals are chrome, iron, coal, nickel, gypsum, sulfur, mercury, asbestos, marble, and salt. There also has been some prospecting for oil, but not much has been found.

Energy

Because the Philippines is dependent on petroleum for energy, every year the country must spend more than $2 billion to import crude oil. The only local energy source that puts a dent in oil needs comes from geothermal power (natural Earth-heated water vapor). Geothermal power provides the annual equivalent of nearly 8 million barrels of oil. The United States is the only country with more geothermal energy than the Philippines.

MANUFACTURING

The manufacturing sector is the largest single contributor to the Philippines' economy, accounting for almost 26 percent of the GDP. Most manufacturing occurs in and around Manila and includes processing of agricultural products, textiles, vehicle components, and electronics.

SERVICE INDUSTRIES

Another important segment of the Philippine economy is the service industries. This includes services such as transportation, communication, and tourism.

Transportation

Because it is an archipelago of thousands of islands, marine travel is an important industry in the Philippines. From pump boats up to large passenger ships, there is a complex industry of water transportation in the Philippines. This is both the oldest and the cheapest mode of transport in the Philippines. Most cargo between the islands travels by boat, and Filipinos make up a big part of the international shipping crews.

Recently, air travel has become developed as a quicker, although much more expensive, way to move among the islands. There are about 260 airports in the Philippines, but only about one-third of them have paved runways.

During their short time in the Philippines, Americans concentrated their efforts on developing the land transportation infrastructure. Now, roads have been extended to nearly every community, and extending road access has become the priority of many mayors of remote towns. There are more than 124,274 miles (200,000 kilometers) of highways in the islands, with only about one-fifth of these highways being paved. Very few Filipino families own cars. Local in-town travel is most common on *jeepneys* (large covered jeeps) or motorized *trikes* (motorcycles with covered sidecars). Longer distance travel is more common on buses.

Communication

Given that the Philippines is a country of thousands of separate islands, new communication technology has boosted its connectedness. In the past few years, Internet services have gone from a luxury of the elite to a readily available convenience for Filipinos anywhere. Even very small and remote towns now have Internet cafés, where the cost of using the Internet is only about 20 pesos (38 cents) an hour. Despite being a "poor" country, the Philippines ranks twenty-first in the world in the number of Internet users.

Another widely popular and quickly growing communications technology is cellular telephones. Just a few years ago, before cell phones, you can imagine what a nightmare it was to have telephone connections among all those islands. It was so

Most Filipinos do not own cars, and some remote towns do not even have road access. In the cities, Filipinos rely on public transportation such as buses or pedicabs, seen here.

expensive and difficult to have landline wire hookups that only the wealthy families and businesses could afford to have a telephone. Most Filipinos were forced to go to calling stations and pay high prices just to call a relative on a nearby island. Even worse, they had to arrange for that relative to be at his or her calling station at the time of the call. All of these hassles have disappeared with the development of cell phones. For the first time, Filipinos of a wide range of economic classes from the cities to the remote provinces are able to communicate. The cost of a phone call can be as little as 4 pesos (7 cents) a minute. Even cheaper and much more widely used is the text-messaging feature on cell phones. This allows people to send short typed messages to each other for a fraction of a penny apiece. It is little wonder that the number of registered cell phone users is growing at a rapid rate. Although cell phones are still fairly new to the country, the number of them in use is already more than double the number of main landlines.

Tourism

In recent years, the Philippines has made a big push to increase the country's attractiveness to foreign tourists. With islands of white sand and coconut trees and cultural sites such as the famous rice terraces of Luzon, the Philippines seems like it would draw many Western tourists. The country has some major drawbacks that Westerners hear about frequently in the international news, though. With its high crime rates, violent demonstrations, and sinking ferries, floods, and other disasters, the Philippines scares off many potential tourists. Even so, the number of tourists visiting the Philippines is slowly increasing and tourism is becoming an increasingly substantial part of the economy. Currently, about 1.5 million tourists visit the Philippines per year.

THE PHILIPPINE ECONOMY AND THE FILIPINO

How the Philippine economy affects average Filipinos can be described by their buying power, how much money they can make, and how much it costs to live. As already mentioned, the buying power of Filipinos is only about one-ninth that of Americans. For this reason, it is true that foreign items usually are not on a Filipino's list of things to buy.

When one considers the Philippine minimum wage, it does seem as if Filipinos do not earn much money at all. Consider the fact that in metro Manila, the minimum wage is equivalent to about $5 (280 pesos) a day at current exchange rates. The minimum wage in the United States is much more than this amount, at $6 per *hour*. That means that after a day of working a minimum wage job, a Filipino in the Philippines would walk away with $5, whereas an American doing the same job in the United States would walk away with $48.

Perhaps making less money is all right if things in your country cost less. The things Filipinos need to live in the Philippines cost much less than they would in the United States. Many rural Filipinos live in thatch houses that cost roughly $200 to build, and Filipinos generally own their houses. In the United States, it is not uncommon for houses to cost 500 times more than this. Obviously, ideas like "poverty" and "wealth" have to be put into perspective to be really understood.

One clear pattern in the Philippines is that average incomes are higher in urban areas than in the provinces. The minimum daily wage outside of metro Manila is generally about 100 pesos less than inside Manila, and average urban family incomes are nearly twice that of rural family incomes. Some of this difference is explained by cost of living: It is more expensive to live in the city than in the provinces.

Another income disparity—between the wealthiest and poorest Filipino citizens—is harder to explain. The top one-third of the wealthiest individuals in the Philippines own more than two-thirds of the country's monetary wealth.

CHAPTER

7

Living in the Philippines Today

The Philippines is hardly the quiet set of individual islands that it was before the arrival of the Spanish. Today, it is a colorful, bustling country, blooming with smiles, opportunities, fiestas, and creativity. After all their struggles together as a nation of islands, Filipinos are learning what it means to be Filipino. There is a much greater sense of nation than in the past, and, with newer and better transportation and communication services, Filipinos are connected with each other more now than ever before. Filipinos today are able to travel throughout the islands for their work or for their vacations. Cell phones and the Internet have accelerated this nationalism by bringing Filipinos together daily for various causes. In this chapter, you will learn a little about what it is like to be Filipino today. You will see how Filipinos live, their opportunities for education and careers, the holidays they celebrate, and what they do with their free time.

Education has important value in the Philippines, and even poor, rural families are committed to ensuring that their children attend some school. These elementary school students sit in a classroom in Panay.

EDUCATION AND CAREER OPPORTUNITIES

The school system in the Philippines is based on the American model. They have a six-year elementary school followed by four years of high school. You may have realized that that adds up to only 10 years of school, which is one way that Filipino schools differ from American programs. The Philippines does not have middle schools or junior highs, so students receive their diplomas two years earlier than students in the United States, at age 16.

Education is a common value throughout the Philippines. Despite being a poor country, the Philippines is considered one of the most-schooled countries in Asia. During American colonization, the United States emphasized the importance of education and invested heavily in the Philippines' educational system. They built schools and brought in American teachers as volunteers. Now, from Manila to the most remote barangays, parents are committed to giving their children a chance to go to school. In rural areas, it is common to see children walking long distances on mountain trails just to get to class. The value Filipinos place on education is reflected today in the fact that the country's literacy rate (the ability to read and write) is more than 95 percent.

Nonetheless, the amount of school that youngsters attend and the quality of their education varies throughout the country. This difference exists especially between urban and provincial school systems. Young children are required to attend at least the first four years of the six-year elementary school. After that, some students, especially in the provinces, have to drop out or attend part time, because their help is needed on the family farm or in the family business. Unfortunately, this happens a lot. In the whole country, only two-thirds of Filipino students complete the entire six years of elementary school and less than half make it all the way through high school. In big cities like Manila, nearly 100 percent of the students finish elementary school. The problem is in provincial areas where people are poorer and there are fewer schools. Only about one-third of the youngsters in some provinces ever finish elementary school.

There are problems with the quality of education these children get, as well. In an international test taken in 1995 by 500,000 elementary and high school students in 45 countries, the Philippines ranked third to last in elementary math and second to last in elementary science. Nearly all Filipinos are basically literate, but other literacy measures better reflect these quality problems. For example, only 87 percent of Filipinos are what is known as "functionally literate;" this refers to the ability to process numbers. Again, the quality of education differs depending on whether you are in the city or the provinces. Private schools, only available in urban areas and to the wealthy, generally provide higher-quality education than do the public schools.

One of the basic problems for elementary schools is that they lack resources. More than half do not have electricity or running water and lack educational supplies such as books or simple things such as tables and chairs. Most of these problems are especially evident in provincial schools. In many rural areas, there are few permanent teachers, so students may go without school for months at a time, especially during the rainy season, when travel is difficult.

Only 14 percent of Filipinos get college degrees (compared with 25 percent of Americans). For those Filipinos who study at colleges and universities, many wish to become doctors, nurses, lawyers, architects, and increasingly, information technology professionals. Many Filipinos excel at these professions. Philippine nurses, for example, are well known for the quality of their care and professionalism. On the other hand, many of the Filipinos who receive higher education end up leaving the Philippines to work in other countries, where salaries may be much higher. This phenomenon, known as "brain drain," is a problem. The skills these Filipinos obtain through education are much needed back home.

SOCIAL SERVICES

Health care is important to Filipinos, as it is elsewhere. Despite its being a poor country, the Philippines' health care generally is considered very good. Many of the medicines that are expensive or difficult to get in the United States are much cheaper and more accessible in the Philippines. The same is true for medical professionals. In the United States, it is unusual to see a doctor for less than $100, but in the Philippines the cost might be closer to $5. Instead of having to drive to the doctor's office, the doctor might even come to your house.

Other social services are handled very differently in the Philippines than in the United States. In the United States, young children often are dropped off at day care while their parents head to work. In the Philippines, it is much more common for a family member, such as a grandmother or older daughter, to take care of the young children of working parents. If there are no relatives at home who can look after children, it is very common to hire a *ya-ya*, who is like a nanny and works for minimal wage.

Taking care of elders is another important social service that Filipinos handle very differently from Americans. It is almost unheard of for elderly people to leave their homes to live in a nursing home, as is common in the United States. Instead, when people become too old to care for themselves, they move back in with their children and continue living as active members of a household.

Another unexpected or unusual difference in Filipinos' day-to-day lives is the number of service workers they might have at home. In addition to ya-yas and someone looking over the elderly, there may be many more people hired to manage daily household affairs. It is common for a family to offer food and board and pay tuition expenses to young girl students who agree to cook and clean for the family. Some Filipinos hire gardeners, who are responsible for keeping the lawn mowed and the plants alive. Also, very few of the people who own cars drive them. Instead, they hire personal drivers, whose job it is to not only drive everywhere but also to keep the car clean and in good repair.

HOLIDAYS AND CELEBRATIONS

Filipinos love parties. With all the public holidays, religious holidays, weddings, baptisms, and town fiestas, there is always a reason to celebrate. Almost every week of the year has an official holiday somewhere in the Philippines. Most of these are Catholic holidays because such a high percentage of Filipinos belong to this religion. Of the Catholic holidays celebrated nationally, you probably recognize both Christmas and Easter. In observance of all the holidays around Easter, Filipinos celebrate for the entire week before Easter Sunday. This week, called Holy Week, is one of the biggest holidays in the country, when all government offices and businesses close and most Filipinos return to their home provinces.

All Saints' Day is another big national Catholic holiday when Filipinos head to the cemeteries to gather at the gravesites of their ancestors. Families usually clean and decorate these graves and bring music and food to have picnics "with" their ancestors. Despite having to do with death and graveyards, this holiday is joyous and the cemeteries end up hosting big parties. In some regions, family members go even further with this tradition. They dig up the bones of their dead relatives and clean them off to take pictures with them. This may sound very gruesome to us, but it is meant as an honor to the ancestors and is thought to bring good fortune to the survivors.

In addition to religious holidays, some of the national holidays recall important historical events. Independence Day (June 12) is when Filipinos gained independence from the Spanish. People Power Day, on February 25, celebrates the large nonviolent demonstration that lead to the over-throw of the powerful Marcos regime. Bonifacio Day, or National Heroes Day (November 30), honors the many national heroes who died for their belief in the Philippine nation. The most famous hero, José Rizal, is remembered on his own holiday on December 30, which is the anniversary of his public assassination.

There are many holidays celebrated regionally that include patron saints and histories or traditions unique to certain provinces in the Philippines. These include Bataan

The Filipino calendar is filled with celebrations, from public holidays, to religious feast days, to local fiestas. Here, children dressed as the patron saint St. Niño parade down a street in Tacloban City during the festival of Kasadyaan, a celebration of the culture and history of the region.

Day commemorating the Bataan Death March in Bataan Province. Bahug-Bahugan, on Mactan Island, is a holiday in which the battle that lead to Magellan's death is acted out in the same place where it happened. Other regional holidays are much lighter in nature, such as Carabao Carroza Race near Iloilo City, where water buffalo are raced through the neighborhoods. The Feast of San Juan Bautista (Saint John the Baptist) is celebrated in San Juan, Manila, by throwing water at friends or even unknown people to symbolize the baptisms of Saint John. Turumba Festival in Laguna Province is celebrated by a parade of falling, jumping, leaping, and dancing people.

More locally, every town in the Philippines has its own celebration once a year, usually called the town fiesta. These may coincide with the celebration times for patron saints of these towns or with other big holidays such as Christmas or New Year's Day. Many of the town fiestas occur in May in the heat of the Philippine summer. These fiestas can be most closely compared to county fairs in the United States, although they seem much bigger. Fiestas usually have many large group celebrations in the town's center. Events include speeches by the mayor and other politicians, contests, dances, beauty pageants, and lots of music. Many vendors arrive for the town fiesta, and colorful markets offering stalls of food, clothing, and games may stay open all night. For Filipinos, town fiestas are like homecoming celebrations, when family members who work far away come together with their extended families. This is a time when every house in the town is lively, overflowing with music, food, and lots of friends and family.

LEISURE TIME

Filipinos tend to be very happy and social people, and they enjoy spending their free time together. In the provinces, leisure activities vary by region. After school, youngsters stroll through the markets, play basketball, or listen to music. In towns with Internet cafes, chatting in chat rooms and playing video games are very popular. Movies are a popular way to enjoy free time.

Adults may also spend leisure time doing any of these as well. Women enjoy playing games like Chinese mah-jongg, cross-stitching, or just hanging out together telling stories and having snacks. Women often combine socializing with their work. A lot of good information gets passed around as the laundry is done at the river. Women in the Philippines do not typically drink alcohol, but men, starting as early as high school, often include alcohol in their social plans. Men socialize in ways similar to how they might in the United States. They play card games, watch sporting events, go hunting, or simply hang out with a guitar and lots of stories.

Leisure time in the cities is spent in much the same way it is spent in the country. Especially in large cities, the most popular gathering place for spending free time is in the large shopping malls. There, friends can go window shopping, get snacks, listen to music, play video games, watch a movie, and much more. Malls have the added benefit of being air conditioned, which may offer a cool break from the heat and pollution outside. Mall life is much more popular in the Philippines than in the United States. In Manila, malls can get so crowded that it is impossible to walk more than two steps without running into someone.

Sports and Recreation

Basketball is easily the national sport. It may seem strange that a sport that relies on tall people would become popular in a country where the average height is much shorter than in the United States. With their height advantages, lots of American and European basketball players end up playing in Filipino professional leagues. Basketball is so popular that even in the most remote areas, there will be a basketball court with hoops on palm trees and lines drawn in the dirt.

Some of the other popular recreational activities are billiards (pool), dancing, and cock fighting. Ballroom dancing, which was popular in the United States in the 1950s, is still very popular in the Philippines. It is expected that everyone knows the basic steps. You can see a different type of dance at a cock fight. Although it is extremely violent and illegal in some countries, including the United States, cock fighting is an extremely popular form of entertainment in the Philippines. Spectators bet on one rooster to win the fight, and the birds are let loose on one another. In a flurry of feathers and snapping beaks, the fight is over in a matter of seconds, with only one rooster the victor and the other one probably dead.

CHAPTER

8

The Philippines Looks Ahead

In the preceding pages, you have learned a lot about the Philippines, a country that has a long history with the United States and yet one that most Americans know very little about. You have read how this archipelago of unrelated islands got grouped together as a colony by a foreign power, and you have learned how, in their collective resistance, the Philippine people found themselves operating as a united nation. Since initiating this nationalistic movement for independence against the Spanish, Filipinos have struggled to create a nation with its own identity. They have continued to develop this identity through more than a half-century as an independent nation. With a functioning democratic government and a role in the global economy, the Philippines now has a foundation on which to build its future as a country. This future no doubt will have its share of important challenges.

One of the first challenges the Philippines will need to face is the mounting pressure on the environment. The depletion of natural resources has resulted in very little to show for the country's development. Poor governance has led to a reduction of these resources without thoughtful reinvestment, so that few benefited at the expense of many. The Philippine people are major victims of this oversight. They are now lacking the forest products, clean water, and clean air that the natural forests used to provide. The poor supply of environmental resources is already evident in the amount of pollution in the Philippines. With its amazing population density, Manila is beginning to choke on its own air pollution, and clean sources of water are scarce during parts of the year. Other victims of the inadequate environmental management are the thousands of unique animal and plant species. This world-famous wealth of biodiversity will have a difficult time surviving in the few remaining and still-shrinking forests of a once great and secure heritage. The country is certain to face other unprecedented environmental challenges in the future as well. Steep projected rises in ocean levels from global climate change, for example, will create hardships for the archipelago that it is little prepared to address.

Another challenge of the future of the Philippines is its rapidly growing population. In addition to its contribution to the Philippines' dwindling natural resources and mounting pollution, the uncontrolled population growth poses many problems. With nearly half of the Philippines' people living below the poverty line, one can only wonder what will happen to all the new people being added to the country. The Philippines already has to resort to imported rice to feed its people; how will it handle double this population in just a couple of decades? Also, the Philippines' infrastructure is insufficient for the current population size, with electrical shortages happening daily. How will it manage to upgrade these facilities to handle double the load with few funds available for these improvements?

The Philippines' government is also likely to experience some growing pains in the future. The centralized democratic government

Today, the Philippines faces many challenges, from pressure on the environment, to poverty, to governmental instability. However, in its short time as an independent country, the Philippines has developed a strong foundation to build its future as a member of the world community. These young Filipinos grin after performing at a 2003 trade exhibit in Manila promoting Philippine cultures.

is still young and transitioning to the type of government most effective for the country. In its short decades of independence, it has made great leaps in developing this government. With one of the highest voter turnout rates in the world, it is clear that Filipinos have faith in their political processes. Also, in overcoming the powerful Marcos dictatorship through a mass nonviolent protest, Filipinos have realized the power of the people. The Philippine government is made up mostly of the privileged elite, but the empowerment of the common Filipino has served as an important check on the government. Filipinos have made it clear that, although a certain amount of poor governance will be tolerated, there is a point at which those in charge will be held accountable for their actions.

Despite its advances, the Philippines must overcome many obstacles before its government will be as functional and stable as people need. Large gaps between the country's different economic classes and religious groups will make it difficult for politicians to appeal successfully to everyone. Also, small-scale corruption runs rampant in the Philippines' governmental units, making these offices much less effective than they should be.

Like the strangling fig trees that wrap around its forest trees, the Philippines' economy has built itself around the strength of supporting foreign economies. Most countries would prefer not to be dependent on other countries, but foreign economic support is very advantageous in the Philippines. With the help of foreign aid through loans and grants, the country slowly is developing the infrastructure it needs to compete in global markets. Also, foreign investment brings in a lot of income to those who otherwise would have no jobs or only low-paying jobs. Computer and information technology jobs, in particular, are becoming abundant with foreign investment. With the influx of these resources in the past 20 years, the Philippines' middle class finally has begun to grow. Middle-class incomes in other countries have been associated with higher education levels and lower birth rates, both of which would strengthen the Philippines' economy.

In many ways, the collection of islands called the Philippines still struggles to operate as one big nation. The roots of most of its environmental, governmental, and economic problems are on the national scale. The Philippines' potential for success, however, definitely lies in the strength of its individual islands and local communities. Strong Filipino family and community ties have maintained a social fabric through the many hardships and identity crises the Philippines has faced. In these tightly knit communities, you can find effective and progressive governance, economies that include and provide for everyone, and efforts toward the sustainable use of the environment. Families' commitment to each other has allowed local communities to withstand the difficulties of a long history of exploitation and will be the strength that encourages them to tackle future problems.

Country Name	Republika ng Pilipinas (Philippines)
National Language	Pilipino (or Filipino)
Capital	Manila
Land Area	114,229 square miles (298,170 square kilometers)
Highest Elevation	9,691 feet (2,954 meters)
Climate	Tropical marine; northeast monsoon (November to April), southwest monsoon (May to October)
Population	84,619,974 (2003 estimate)
Population Growth Rate	2% per year
Life Expectancy	69 years
Infant Mortality	25 deaths per 1,000 births
Ethnic Groups	Christian Malay, 92%; Muslim Malay, 5%; Chinese, 1%; other, 3%
Religions	Roman Catholic, 83%; Protestant, 9% Muslim, 5%; Buddhist and other, 3%
Literacy	96%
Government	Constitutional democracy
Independence	June 12, 1898 (from Spain); July 4, 1946 (from the United States)
Currency	Philippine peso
Gross Domestic Product	$379.7 billion (2002 estimate)
Unemployment	10%
Exports	Electronic equipment, machinery and transport equipment, garments, coconut products, chemicals
Imports	Raw materials, machinery and equipment, fuels, chemicals

History at a Glance

110,000 B.C.– 50,000 B.C.	The first people arrive over land bridges for hunting.
25,000 B.C.	Aeta people arrive over land bridges and settle.
1500 B.C.– 500 A.D.	Malay immigrants begin arriving by boat.
500–1500	Five waves of Malay immigrants arrive by boat.
800–1478	Buddhist-Hindu culture is integrated into the Philippines by Indian immigrants from Indonesia.
960–1279	Chinese merchants come to the Philippines from Indochina to trade porcelain crafts for native wood and gold.
1380	Arab Muslim scholar Makdum arrives in the Sulu Islands to propagate Islam.
1475	Sharif Mohammed Kabungsuwan establishes an Islamic center in the Sulu Islands.
1521	Ferdinand Magellan arrives and claims the Philippines for Spain; he is killed on Mactan Island by Chief Lapu Lapu.
1543	The islands are named "Filipinas" after King Philip II of Spain.
1896	José Rizal is executed by firing squad in Manila for pushing for Philippine independence from Spain.
1898	The Americans win the Spanish-American War, freeing the Philippines from Spanish rule.
1941	Filipinos join the Americans to fight against the Japanese in the Philippines in World War II; 10,000 Americans and Filipinos die at the hands of Japanese during the gruesome Bataan Death March.
1946	At the end of WWII, the Americans grant full independence to the Philippines.
1972	Ferdinand Marcos declares martial law over the Philippines.
1983	Ninoy Aquino is assassinated after returning to the Philippines.
1986	The People Power march ousts Ferdinand Marcos and puts Cory Aquino in power as president.

1991 Mount Pinatubo erupts, burying thousands of homes under nine feet (three meters) of ash.

1992 The American military leaves the Philippines for the first time in nearly 100 years.

2000 People Power II march overthrows Joseph "Erap" Estrada and gives presidential power to Gloria Macapagal Arroyo.

Further Reading

Brainard, Cecilia M. (Editor) *Growing Up Filipino: Stories for Young Adults* (Philippine American Literary House, 2003).

Broad, Robin, John Cavanagh, and Robert Broad. *Plundering Paradise: The Struggle for the Environment in the Philippines* (University of California Press, 1993).

Gritzner, Charles F. "Third World Peoples and Problems: A Cycle of Frustration," in Gail L. Hobbs (ed.), *The Essence of PLACE: Geography in the K-12 Curriculum* (Los Angeles: California Geography Alliance, Center for Academic Inter-Institutional Programs, UCLA, 1987; pp. 301–313).

Guillermo, Artemio, R., Tina Sevilla (Illustrator), Nimfa M. Rodeheaver. *Tales from the 7,000 Isles: Popular Philippine Folktales* (Vision Books International, 1997).

Hamilton-Paterson, James. *America's Boy: A Century of Colonialism in the Philippines* (Henry Holt & Company, Inc., 1999).

Karnow, Stanley. *In Our Image: America's Empire in the Philippines.* (Ballantine Books, 1990).

Langellier, John P. *Uncle Sam's Little Wars: The Spanish-American War, Philippine Insurrection, and Boxer Rebellion, 1898–1902 (The G.I. Series)* (Chelsea House Publishing, 2001).

McReynolds, Patricia J. *Almost Americans: A Quest for Dignity* (Red Crane Books, 1997).

Nickles, Greg. *Philippines: The Culture* (Bt Bound, 2002).

Peters, Jens. *Lonely Planet Philippines (6th Ed)* (Lonely Planet, 6th Edition, 1997).

Roces, Alfredo, Grace Roces, Shirley Eu (Illustrator). *Philippines (Culture Shock!)* (Graphic Arts Center Publishing Company, 2002).

Websites

**Conservation International's biodiversity hotspots
Philippines information page**
http://www.biodiversityhotspots.org/xp/Hotspots/philippines

Geography Home Page
(General information about many topics including the Philippines)
www.geography.about.com

"Philippines," *CIA World Factbook*
(annual editions) at:
http://www.cia.gov/cia/publications/factbook

"Philippines," U.S. Department of State Country Profiles
(annual updates) at:
www.state.gov/r/pa/ei/bgn

"Philippines-A Country Study,"
Library of Congress, Federal Research Division,
Country Studies (regularly updated) at:
http://lcweb2.loc.gov/frd/cs/petoc.html

"Vanishing Treasures of the Philippines Rainforest"
(Current information about thenatural environment
of the Philippines and its unique wildlife species.)
http://www.fmnh.org/vanishing_treasures/

"WOW Philippines" Homepage
Official tourist information homepage.
www.tourism.gov.ph

Index

Index

Index

Index

TAMMY MILDENSTEIN is a Doctoral Student in the Wildlife Biology Program at the University of Montana, in Missoula. She lived in the Philippines with her husband (the other author of this book) for four years as a U.S. Peace Corps volunteer. There, Tammy completed her master's degree on the habitat use of threatened Philippine flying-foxes and began her career in wildlife biology and conservation. Her work on the conservation of Philippine flying-foxes and Philippine forests has continued to the present day and is the focus of her Ph.D. research. Her Bat Count project includes the help of Filipino biologists throughout the islands and has lead to the creation of the Philippines' first nationwide bat monitoring program. Bat Count has won many awards including the Gold Award from the British Petroleum Wildlife Conservation Programme. When Tammy is not working in the Philippines, she lives in Montana, where she raises bees, tends her garden, and spends as much time as possible outside watching wildlife.

SAMUEL CORD STIER is a Doctoral Student in the Department of Forestry and Conservation at the University of Montana. He completed his master's degree on the dietary habits of threatened wildlife species in the Philippines, where he lived and worked as a U.S. Peace Corps Volunteer from 1997–2001. Sam's ecological and policy work has centered on the restoration of forests in the tropics. In addition to initiating one of the first habitat-based forest restoration projects in the tropics (at Subic Bay, Philippines), he has worked with several environmental and development organizations on facilitating and improving tropical forest restoration practices. With his wife Tammy Mildenstein, he lives in western Montana in a house by a creek next to the Rattlesnake Wilderness, where he loves to go hiking.

CHARLES F. ("FRITZ") GRITZNER is Distinguished Professor of Geography at South Dakota University in Brookings. He is now in his fifth decade of college teaching and research. During his career, he has taught more than 60 different courses, spanning the fields of physical, cultural, and regional geography. In addition to his teaching, he enjoys writing, working with teachers, and sharing his love for geography with students. As consulting editor for the MODERN WORLD NATIONS series, he has a wonderful opportunity to combine each of these "hobbies." Fritz has served as both President and Executive Director of the National Council for Geographic Education and has received the Council's highest honor, the George J. Miller Award for Distinguished Service. In March 2004, he won the Distinguished Teaching award from the American Association of Geographers at their annual meeting held in Philadelphia.